Arzygul Yuldashevna KIDIRBAEVA

ECOLOGY OF THE WOLF (CANIS LUPUS LINNAEUS) IN THE SOUTHERN ARAL SEA REGION

Arzygul Yuldashevna KIDIRBAEVA

ECOLOGY OF THE WOLF (CANIS LUPUS LINNAEUS) IN THE SOUTHERN ARAL SEA REGION

ScienciaScripts

Imprint

Any brand names and product names mentioned in this book are subject to trademark, brand or patent protection and are trademarks or registered trademarks of their respective holders. The use of brand names, product names, common names, trade names, product descriptions etc. even without a particular marking in this work is in no way to be construed to mean that such names may be regarded as unrestricted in respect of trademark and brand protection legislation and could thus be used by anyone.

Cover image: www.ingimage.com

This book is a translation from the original published under ISBN 978-620-7-45769-4.

Publisher:
Sciencia Scripts
is a trademark of
Dodo Books Indian Ocean Ltd. and OmniScriptum S.R.L publishing group

120 High Road, East Finchley, London, N2 9ED, United Kingdom
Str. Armeneasca 28/1, office 1, Chisinau MD-2012, Republic of Moldova, Europe
Managing Directors: Ieva Konstantinova, Victoria Ursu
info@omniscriptum.com

Printed at: see last page
ISBN: 978-620-3-48872-2

Contents

Based on many years of research, the book considers ecological features and spatial and territorial distribution, population dynamics, sex and age structure, ecological and economic values of the wolf population in ecosystems of the Southern Priaralie. The monograph is intended for biology and ecology students, hunting specialists, employees of specially protected natural areas and a wide range of readers interested in wildlife.

K.M. Atanazarov - Candidate of Biological Sciences, Associate Professor of Nukus branch of Samarkand Institute of Veterinary Medicine.

A.I. Kurbanova - Candidate of Biological Sciences, Associate Professor at Berdakh Karakalpak State University.

INTRODUCTION

The increase in population and the development of virgin lands has led to a reduction in natural resources, including a decrease in the diversity of wildlife resources and the disappearance of populations. Especially against the background of urbanisation, the increased pressure of hunting has caused the reduction of natural habitats and negative changes in the structure of the range of predatory animals. In this regard, the determination of the role of predatory animals in the biocenosis and the management of their natural populations are of great scientific and practical importance. In the world in certain urbanised areas, special attention is paid to studies of individual characteristics of predatory animal populations, assessment of their ecological structure and distribution range, and substantiation of ecological features of their territorial groupings.

Particularly in areas where anthropogenic impact and climatic changes are strongly pronounced, elucidation of changes in predator populations in narrow ranges makes it possible to explain the ecological mechanisms of predator-prey relationships and to use them rationally in practice.

The South Prearalie region is not only one of the centres of animal biodiversity in Central Asia, but also one of the key regions rich in natural populations of predatory mammals with a narrow range structure, including the wolf *(Canis lupus* Linnaeus). Determination of the distribution area and ecological structure of the wolf population in all natural zones of the Southern Priaralie, elucidation of the dynamics of wolf numbers under the influence of external factors, as well as the development of effective methods of regulation acquire actual scientific and practical importance.

At present in the republic special attention is paid to protection of fauna and rational use of resources. On the basis of the programme activities carried out in this direction certain results have been achieved, in particular in the field of preservation of biodiversity of the Aral region, protection of resources of hunting species of wild animals and increase of their numbers. The strategy of action for the further development of the Republic of Uzbekistan[1] set out the tasks for "....global warming and mitigation of the consequences of the Aral Sea tragedy". Based on these tasks, in particular, determination of the wolf habitat area in the conditions of the Southern Priaralie, identification of the peculiarities of feeding ecology, disclosure of age and sex structure of the population and on this basis development of recommendations on management and regulation of the wolf population acquire important scientific and practical importance.

In the Republic of Uzbekistan, studies on the ecology of wolves, their distribution in ecosystems and their economic importance were given in the works of Palvanizov M. (1974, 1990), Nuratdincv T. (1969), Reimov R. (2000, 2003). But the above-mentioned data cannot provide complete information on the ecology of the wolf *(Canis lupus* Linnaeus) in the conditions of the Southern Priaralie. Therefore,

---

[1] Decree of the President of the Republic of Uzbekistan No. UP-4947 of 7 February 2017 "On the Strategy of Actions for the Further Development of the Republic of Uzbekistan".

justification of the study of wolf ecology, population counting and monitoring in the conditions of the Southern Priaralie are of theoretical and practical importance.

In the course of the study, the ecological features and spatial-territorial distribution of the wolf were determined for the first time in the conditions of the Southern Priaralie, the population dynamics and sex-age structure were established, and the ecological and economic significance of the wolf in the ecosystems of the Priaralie was revealed. The accounting of wolf abundance in indigenous areas was carried out and differentiated ecological and geographical principles for regulating their numbers were developed.

CHAPTER I

The study of the **ecology of predatory** mammals in different regions of the world is of great scientific **and** practical importance. In recent years, the question of the role and importance of predators in nature has caused active discussions, where the role of this or that predator was interpreted from subjective positions, without involving actual scientific material [18; p.123-193, 75; p.8-50, 8; p.605, 141; p.7-95]. The predatory lifestyle in higher vertebrates a priori implies an active way of life support, an influential biocenotic role, direct intensive impact on other members of the community, and a high level of ecological and social risks. The study of ecological and ethological aspects of predation is of key importance for solving many questions of modern theoretical and applied population ecology.

In population ecology, the study of predation on specific species (in particular the wolf) is necessary to understand the factors and mechanisms of spatial and demographic dynamics of populations and functional relationships in communities and ecosystems. This knowledge is essential for the development of animal conservation measures, population management, and the recovery of endangered and low abundance species of mammalian predators.

There have been quite a lot of researches on the study of wolf ecology. Thus, in the European part of Russia and neighbouring territories, publications devoted to wolf ecology were published by Smirnov V.S. (1985), Korytin S.A. (1976), Zyryanov V.A. (1995), Lineitsev S.N. (1983), Bibikov D.I. et al, (1985), Bondarev A.Y. (2002, 2012), Pavlov M.P. (1990), Fedosenko A.K. (1986), Badridze Y.K. (2004), Suvorov A.P. (2009) and others.

Original observations on the ecology of forest-steppe wolves in the south of Krasnoyarsk Krai and Khakassia are contained in the works of Kozlov V.V. (1966), materials on the ecology of Sayan mountain taiga wolves in the works of Zavatsky B.P. (1986,1987), Smirnov M.N. (1984), Siberian wolves in the works of Suvorov A.P. (2004). General data on the ecology, distribution and abundance of wolves in Krasnoyarsk Krai (Russia) are contained in the works of Bondarev A.Ya. - (1983, 2002, 2012), Tsyndyzhapova S. V. (2000, 2003) for the Irkutsk region of the Russian Federation.

In 1985, under the editorship of Prof. D.I. Bibikov, the monograph "The Wolf" was published, which contains the results of fundamental studies of wolf populations by many scientists from Russia, the Baltic Republics, Transcaucasia and Central Asia. The history of the wolf origin is described, systematics, habitat, morphology, behaviour, nutrition, reproduction and population structure, numbers and peculiarities of the way of life in different regions of the CIS and adjacent territories are given.

Scientific studies by Bondarev A.Ya. (2002, 2012) cover various aspects of wolf existence, including the issues of population dynamics and areal changes, group structure of the population, peculiarities of hunting and social behaviour of relationships with victims (ungulates). He also describes the complex relationship of wolves with the peculiarities of economic development of landscapes in the region.

In the monograph by Badridze Y.K. (2004) "Wolf. Issues of behavioural ontogenesis, problems and method of reintroduction" describes problems and methods of preparation for reintroduction of captive-bred large predatory mammals on the example of wolf. The necessary conditions of keeping animals in the process of postnatal ontogenesis are shown for the formation of behaviour, including the ability to reason, within the limits of the norm characteristic of the species.

Fedosenko A.K. in his book "Wolves" (1986) gave information about wolf encounters on the territory of Kazakhstan. The occupied place of wolf in zoological nomenclature was described, data on wolf population records were given. Also, special attention was paid to ecological features of wolf habitat and refuges, diurnal lifestyle, gregariousness and social behaviour, reproduction of wolf population in Kazakhstan. The economic and ecological importance of the wolf was noted.

In his works Tirronen K.F. (2009) presents the results of studying the regularities of population dynamics, distribution, ecological and spatial structure of populations and some regional peculiarities of ecology of large predatory mammals (including the wolf) inhabiting the Karelian-Murmansk region.

Shkvyr M.G. (2008, 2012) focused on the assessment of the dynamics and structure of the wolf range and ecological and ethological features of the wolf population in Ukraine. He analysed the current structure and dynamics of the wolf range, ecological and ethological differences of wolf territorial groupings in all natural zones of Ukraine. He also touched upon the multifaceted problem of human-predator conflict, the problem of coexistence of large predators, including the wolf in a transformed habitat.

The scientific studies of Tsyndyzhapova S.D. (2000, 2003) for the first time in the Baikal region provided data on the study of wolf ecology, also determined the systematic status of the wolf, analysed population dynamics, spatial distribution and diet structure.

The scientific research of Hernández-Blanco J. A. (2004) focuses on the spatial-ethological organisation of minimal

population groups of the wolf *Canis lupus* L. in a comparative aspect on the territory of the Voronezh and Kaluga Zaseki Nature Reserves, as well as the Oryol Polessye National Park (Russia). The author studied family plots of minimal population groups of the wolf (*Canis lupus lupus* L.) based on the nature and dynamics of spatial distribution of their individuals during the life cycle.

The publication by Ozolnis J. et al. (2009) shows changes in the nutrition, demographic structure and reproduction of the wolf in Latvia associated with the recent implementation of conservation policies.

A.M. Kudaktin (2009) noted three peaks of fluctuation and pulsation of the wolf range in the Russian Caucasus, and characterised the formed wolf packs by feeding type. The author presented a scheme of management strategies for wolf populations in the Caucasus.

The work of French scientists Becker L. et al. (2009) studied the impact of the wolf on human activities and showed the limits of regulation of the wolf population. In the work of Polish specialists Hedrzak M. et al. (2009) a parametric, qualitative and quantitative analysis of the damage caused by wolves to farm animals in South-Eastern Poland was carried out.

Suvorov A. (2004, 2009) highlighted the main aspects of the issue of wolf population management in Russia. Data on the damage caused by wolves to livestock, as well as methods of population regulation and management strategy of wolf populations were presented. He identified six geographic wolf populations of mixed ecological and management significance within natural geographic zones. The author notes that wolves of these populations differ markedly in their lifestyle, ecological and economic significance, which is important for monitoring the state of predator resources, differentiated management and targeted spending of finances to reduce the number of the most "harmful" predators for humans.

Boreyko V. E. in his book (2011) provides legal, philosophical and ethical and ecological arguments in defence of wolves.

The monograph "Biological Signal Field" discusses the problems reflecting the development of Naumov N.P.'s concept of biological signal field over the last 30 years [87, p.323]. They concern methodology, methods of research and description of the biological signal field, including quantitative methods, as well as sensory mechanisms of perception by animals of information transmitted through its channels.

The book "Mammals of the Soviet Union" by Geptner V.G. et al. (1967) "Mammals of the Soviet Union" describes the biology, systematics and economic importance of the land predatory mammals, including the wolf. Works by S.T. Korytin (1976, 2010, 2011) are devoted to the study of the problem of transmission and reception of information in the animal world, the "language" of smells, chemical orientation and signalling of mammals, the use of these properties of animals to control their behaviour and regulate their numbers.

The monograph by Tumanov L.L. (2003) "Biological peculiarities of predatory mammals in Russia" presents long-term materials of field and experimental studies in Russia. The issues of food specialisation and seasonal change of food, daily activity and movement of animals, timing of mating and fecundity, development and speed of young wolf population are covered in detail. Special attention is paid to ecological, physiological and morphological adaptations of organ systems, as well as the functional state of predators (including the wolf) and their populations from the level of helminth infestation.

An annotated phylogenetic system of modern mammals in the volume of the world fauna is presented in I.Y. Pavlinov's book "Modern Systematics of Mammals" (2003), where 30 orders, 147 families, about1180 genera and a little more than 5000 species are given.

The famous scientist, a specialist in the study of the wolf Pavlov M.P. (1990) in his book described the biology, distribution and habits of the wolf, about the relationship of man, to which is far from unambiguous. The importance of the wolf in nature and national economy is shown. The methods of commercial and amateur wolf hunting are described.

The article by Mozgovoy D.P., Vladimirov E.D. (2002) "Signal fields and animal behaviour in signal-information environment" is devoted to ecological problems of mammalian behaviour, which is based on information received by animals through habitat characteristics. Thus, in their opinion, movement elements, behavioural reactions of the same motivation and parameters of the signal field, which represents the signal-information environment of an animal, can have a numerical expression and are calculable depending on the research tasks. The formalisation of mammalian activity in the sign environment implies complex accounting of three parameters: form, intensity and duration of behaviour.

Scientific studies by Kochetkov V.V. (1988, 2013) are devoted to the study of the influence of hunting pressure on the main population indicators of the wolf in the phase of its population growth. The works show the stability of intrapopulation adaptations of this predator to anthropogenic impact. Attention is emphasised on the plasticity of family territories during the period of population increase, which, in his opinion, contributed to the intensive formation of new families and compaction of the spatial structure of the wolf.

The book by Cherenkov S.E., Poyarkov AD. (2003) is devoted to the study of wolf ecology, the history and current state of their fishery, legislative base and hunting methods, as well as processing and evaluation of hunting trophies are considered.

In the publication of Makridin V. P. et al. (1978) covered the biology of large predators, including the wolf. The data on their numbers and distribution are given, the interrelation of animals with the environment and their economic importance are shown, and specific recommendations for species protection are given.

The scientific studies by Palvanizov M. (1974, 1990) summarise the results of many years of field research on the ecology of predatory animals in the deserts of Central Asia. They contain data on systematics and conditions of existence, distribution of wolf population in biotopes, seasonal phenomena in life of predatory animals (including wolf), extensive data on density dynamics by years, seasonal movements, visits, daily activity, behaviour, nutrition, reproduction, enemies, diseases and parasites of predators, their importance in agriculture and hunting. Also in his scientific works the problems of the system "predator-prey" are considered (forage relations, competitive relations between hunters, vicariate, hunting methods, the role of predators in regulating the number of animals). Much attention is paid to the issues of protection and reproduction of predatory animals, proper organisation of fur harvesting.

Scientific works of Reimov R. (1985, 2000, 2003, 2005, 2006) are devoted to the issues of studying the composition of fauna and the state of populations of mammals, including the wolf, and the regularities of their distribution. The ecological peculiarities of the wolf, ways of their adaptation to the conditions of the Aral Sea region were characterised. The regularity of formation of faunistic complexes of predatory mammals has been clarified, ways of protection and rational use of some species of mammals, including the wolf in the national economy of the Republic of Karakalpakstan have been developed.

Thus, the scientific works of many domestic and foreign scientists contain the results of the study of the pack and territorial movement of wolves in different natural conditions. The given research data showed that the wolf population consists of individuals united in a pack, using certain territories of indigenous areas and single animals, not included in the pack, characterised by high mobility. The size of a habitat area and the density of a single pack were found to depend on food availability, the presence of shelters, the degree of predation by humans and the proximity of water sources. The pack size and territorial movements of the wolf population in different territories play an important role in the stability of life support and their numbers [75, pp. 8-50; 8, p. 605; 141, pp. 7-95; 135, p. 25; 116, pp. 18-22].

Considering the problems of wolf population dynamics, many specialists paid much attention to the impact of various environmental factors on their vital activity. Thus, the scientific publications of most scientists show information on the assessment of wolf abundance in different habitat conditions [93, p. 320; 94, p.76; 19, p.29; 128, p.161; 9, p.4-5]. According to the authors, it is established that the effectiveness of various environmental factors as a mechanism regulating population size in ecosystems depends on the influence of anthropogenic impact in natural conditions. They note that these processes are manifested depending on forage resources and protective conditions, in which intrapopulation mechanisms of population size change function and are regulated by predators themselves and their competitors. The dynamics of animal numbers is also characterised by changes in the sex composition and age structure to periodically changing habitat conditions [93, p. 320; 94, p.76; 9, p.4-5; 114, p.27; 32, p.1-5; 34, p.500; 34, p.64, 132, p.72].

Counting wolf numbers is the most important way to specifically analyse the current state of the population. Among the existing methods of wolf resources estimation, the most reliable is considered to be counting by the method of mapping of wolf habitat areas. Many scientists in their works have presented the results of their research according to the above method [115, p.75; 123, p.192; 98, pp. 131-158; 130, p.20; 131, p.28; 132, p.72; 115, pp.18-22].

Behaviour is one of the most important ways of active adaptation of predators to a variety of environmental conditions. It ensures the survival and successful reproduction of both the individual and the species as a whole. Information on the behaviour of animals is necessary for understanding their ecology (peculiarities of their lifestyle), which, in turn, contributes to the development of problems of nature protection and rational nature management. Many studies by domestic and foreign scientists have presented data on the social organisation and behaviour of animals, including the wolf [24, p.111; 18, p.123-193; 34, p.500; 40, p.90-102; 123, p.192; 8, p.605; 39, p.105; 91, p.351; 25, p.320; 130, p.20; 134, p.25; 121, p.12; 145, p.3-5; 147, p. 834; 153, p. 210].

Nutrition is also one of the most important conditions in the life activity of predatory mammals, including the wolf. Studies conducted by many scientists show that the nature of nutrition is determined by the attitude of a given species to the sources of necessary food substances and determines the position of this animal in biocenoses - natural groupings of organisms [93, p.326; 34, p.500; 8, p. 605; 91, p.351; 140, p.795; 45, p.41-43]. Many authors have conducted studies to determine the current range of wolf distribution in different ecosystems [113, p.1-3; 43, p.26-34; 27, p.165-167; 12, p.176].

To date, unfortunately, the intraspecific system of the common wolf (*Canis lupus*, L.) has not yet been satisfactorily developed. It is presented in full in the works of only some foreign authors [154, p.647-686; 150, p.320-321; 149, p.6; 8, p.605; 92, p.196-197, 3, p.70-71].

Among specialists, the more widespread point of view is that all taxa with modern five species of *Canis (lupus, latrans, familiaris, rufus, hodofilax)* are included in one nominative subgenus *Canis s. str.* [8, p.606; 92, p.196-197; 152, p.483-485; 12, p.176]. Different authors' ideas about the taxonomic boundaries of the "wolf group" are sufficiently similar in the sense that it did not include the species occupying species known to be far from the central position of *Canis lupus*. The intraspecific systematics of the wolf has also been considered in various regional summaries [85, pp. 187-263; 18, pp. 123-193; 92, pp. 196-197; 12, p. 176].

In the studies of American specialists 20 subspecies of the wolf according to the system proposed by Goldman (Goldman E.) in 1944 are indicated. On the European territory up to 12-15 subspecies are allocated, and on the territory of the CIS countries the system including 9 subspecies is allocated [18, p.123-193].

For the entire Palaearctic, 12 subspecies were identified according to the data of scientists Ellerman and Morison Scott (1966). In the territory of the CIS countries, four subspecies of the common wolf are indicated [85, p.187-263].

Many works of famous scientists [133, p.487; 138, p.856; 21c.40; 21, p.100; 135, p.568; 76, p.520; 35, p.127; 36, p.64; 39, p.105; 93, p.326; 25, p.320] are devoted to animal ethology. In their opinion, the study of the spatial structure of the population and the nature of space use by individuals and family groups is extremely important in the study of behavioural peculiarities of the ecology of predatory mammals.

The works of some foreign scientists reflect long-term observations of a specific group of wolves, which allow not only to determine the boundaries of the family plot and reveal its internal structure, but also to assess the changes occurring in it [2, p.34; 134, p.25]. Many researchers have provided data on the analysis of the dynamics of the family plot of a group of wolves and the analysis of marking activity of the family group of wolves (*Canis lupus*): the nature of placement of their tags and the contribution of different individuals to the marking of the habitat area [2, p.34; 134, p.25].

Crossbreeding between wolves and dogs is a phenomenon that has long been known in science. However, it is poorly studied and is therefore of great interest. Most scientific articles were devoted to the problems of wolf-dog hybrids [18, p.123-193; 93, p.320; 94, p.76; 8, p.605]. According to their data, black, white, peg legged "outbreds", as well as wolves with intensely red colouring were found in some regions of our country and abroad. The main reason for the appearance of wolf-dog hybrids in nature was a significant decrease in the number of wolves as a result of human persecution, which was accompanied by the disintegration of wolf packs, disruption of the sexual structure of predator populations

Currently, there is a large number of scientific works devoted to the study of the features of dens and shelters of wolves in different regions. Specialists note that for normal existence and reproduction of wild animals it is necessary to have natural shelters or places suitable for family formation and raising offspring. Lack of shelters and absence of conditions for brood nests not only reduce the number of animals, but also lead to their complete extinction in a given area [93, p.326; 94, p.76; 75, p.8-50; 8, p.605; 140, p.7-95; 137, p.72-84].

Many specialists have paid special attention to the study of wolf reproduction processes and identification of generative mechanisms for maintaining homeostasis of this predator's population [18, p.123-193; 150, p.320-321; 93, p.326; 94, p.76; 8, p.605; 91, p.351]. Breeding stages in different geographical populations differ markedly in timing and intensity. According to numerous observations, it has been established that she-wolves of the same pack have heat at different times (early and late). It is well known that wolves (males and females) reach physiological puberty at the age of more than two years [18, p.123-193; 150, p.320-321; 93, p.326; 94, p.76; 8, p.605; 91, p.351; 122, p.52-57; 140, p.7-95,130, p.20; 114, p.24; 45, p.41-43; 47, p.38-41; 48, p.80-81; 27, p.165-167].

Until recently, the issue of wolf population management was not as pressing. From the 1930s onwards, scientists began to express their opinion about revising the entrenched views on the harmfulness of the wolf. In the second half of the 20th century, human attitudes towards nature and biodiversity conservation changed in a certain way. Many scientific publications consider the issues of regulation and management of wolf populations in connection with the large anthropogenic impact on ecosystems [150, p 320-321;93, p.320;94, p.76;8, p.606;144, p.1068-1081;101, p.9-11;82, p.10-12; 131, p. 28;116, p.12-14;38, p.10-15].

One of the fundamental problems of population ecology, which is of great theoretical and economic importance, is the predator-prey relationship. There are no absolutely useful or absolutely harmful predatory animals in nature. Wolves are known to have a certain ecological significance, i.e. by their predation they maintain the number of victims in an optimal state and can serve as a significant regulating ecological mechanism [93, p.320; 94, p.76; 8, p.605; 144, p.1068-1081; 100, p.9-11; 130, p.20; 82, p.10-12; 113, p.1-3; 116, p.13; 132, p.72; 38, p.10-15; 32, p.1-5].

Thus, the generalised literature review has shown that currently the issue of the impact of anthropogenic transformation of ecosystems on the ecological and population features of the wolf population, and in particular, in the region of the Southern

Priaralie, remains insufficiently studied. In this regard, the study of ecological features of the wolf population is of key importance for solving many issues of modern theoretical and applied population ecology.

CHAPTER II

MATERIAL AND METHODS ON WOLF POPULATION ECOLOGY 2.1. Materials, methods and scope of the study

This paper is based on the data collected for the period from 2009 to 2017 in the South Prearalie region on the example of the Republic of Karakalpakstan.

Ground studies were conducted in accordance with the approved programme for the study of ecological features of the wolf population. Field expeditions were carried out in the following districts of Karakalpakstan: Kegeyli, Chimbay, Takhtakupyr, Muynak, Kanlykul and Shumanay, Ellikkala, Beruni and Turtkul districts, as well as Ustyurt plateau and Kyzylkum desert.

In the course of the study we selected model sites located in different territories with different nature management regime. Two model flocks were selected - Aspantai - Shakamanskaya (Kegeyli and Chimbay districts), Uchsai (Muynak district).

In accordance with the research programme, the total length of the ground routes covered was more than 15,000 km. Data on wolf records of the Karakalpak branch of the Sports Society of Hunters and Fishermen of Uzbekistan, Kungrad and Kazakhdarya state forestry farms of the Republic of Karakalpakstan were used.

Common methods were used: itinerary mapping of territorial areas of wolf habitat by means of interviews and questionnaires; identification of individuals by measurement of footprints; analysis of excrement and prey remains; finding and inspection of lairs and dens [85, p.187-263; 98, p.131-158].

The main principle of route counting, mapping of territorial areas of wolf habitat with the help of surveys and questionnaires is to accumulate more observations - meetings with wolves and traces of their activity. Plotting the observation points on the map scheme allows to detect the densification of points - the centres of wolf activity. The more points, the more precisely these centres are determined. By analysing the observations themselves, the season and nature of wolves' stay in the given area are established, and then the total wolf population is calculated for the whole territory under study. Following the principles of the "Guidelines for wolf counting by habitat mapping method" [19, p.29], we used the following methodology [19, p.29] we used the following rules:

1. Mapping all groups of observations in a uniform order.
2. Assessing the validity of observations.
3. Identification of territories with different completeness of accounting.
4. Consistent passage of all information in the prescribed order from correspondents to the controlling organisation.

This method provides high accuracy in determining the number and status of the wolf population. It implies obligatory collection of all information about the wolf. At the same time with the counting of numbers, material is accumulated for solving practical and scientific-practical tasks, of which the following are the main ones:

1. Determination of the magnitude and nature of damage caused by wolf to agriculture and hunting farms.

2. Obtaining a cartographic and ecological basis for wolf population regulation.

3. Establishment of a centralised information bank on wolf ecology.

Thus, the collected material allowed us to develop ecologically competent strategies and tactics for regulating wolf numbers both in conditions of unacceptably high numbers and sparse wolf populations.

At the beginning of the wolf habitat mapping phase, maps of each district and forestry scheme were used.

Based on the questionnaire and survey method, information was collected on the occurrence of wolves in the lands of the region or in a certain area at different times of the year, cases of wolf broods, traces of life activity (excrement, urine point, remains of victims, den, etc.) and cases of wolf attacks on livestock, etc.

The questionnaires were filled in by residents of settlements, hunters, shepherds, gamekeepers, inspectors and other persons by private observations. Any fact, any report related to wolves (including their absence) was considered as an observation. Each received report was assigned a number, which was recorded in the logbook. Then the reliability of the received information was assessed using a five-point system.

5 points - the fact is established by the researcher himself.

4 points - fact checked by the researcher himself or by a knowledgeable, fully credible person;

3 points - the fact is established or verified (or the reliability of information is vouched for) by an experienced gamekeeper, hunter or a person without special knowledge, but quite reliable; in this case the fact does not seem unlikely;

2 points - the fact is established by random, little-known persons, there is a possibility to find an eyewitness, but the verification is not made;

1 point - it is not known by whom the fact was established or the source is obviously unreliable. Scores were given in brackets at the end of each questionnaire.

On the basis of the obtained information, the exact indigenous area of one wolf pack was established and transferred to the map-scheme (Fig. 1). The place of observation was marked on the map-scheme with a point, next to which a circle with a number corresponding to the given observation in the logbook is drawn. The trail, as far as the scale allows, is drawn on the scheme with a line, in the form corresponding to the legacy, and the number of the observation (in this case, the trail) is placed in the circle, at the starting point. The direction of wolves' movement is indicated by an arrow.

At the final stage, a map-scheme was drawn up for the entire study area, which was divided into zones depending on the quantity and quality of information obtained (Fig. 2).

I. The zone of continuous registration. It should include reserves, wildlife sanctuaries and hunting farms with full-time gamekeepers; areas well provided with correspondents; areas assigned to full-time hunters and intensively used by them. The estimated error is no more than 10% of the wolf population.

II. Zone of sufficient information. It includes areas regularly surveyed by active correspondents in most settlements. The possible error is estimated at no more than

20%.

III. An area of insufficient information.

IV. An area of almost complete lack of information.

All of the above zones were labelled with large Roman numerals I-IV. The estimated boundaries of family and pack areas of the wolf population were plotted with dashed lines. Each designated wolf population pack was conventionally labelled with area names. In addition, data on wolf attacks on domestic and wild animals were collected. The facts of attacks on humans were also taken into account in a special way. All found wolf dens, temporary dens, wolf playgrounds were marked on the maps. The accuracy of wolf habitat determination by mapping was guaranteed by as many observations as possible, relating to different seasons and different aspects of wolf lifestyle.

The mapping took into account the locations of direct observations of wolves, their tracks, faeces, scrapes, and information on wolf-dog hybrids. Retrospective data on wolves from the entire study area were also used in mapping.

As a result, a final map-scheme with generalised complete information was obtained. The boundaries of the habitat areas of conditionally labelled wolf populations for which the breeding area is known were determined (Fig. 3). The Aspantai - Shakamanskaya population site served as a model site, for which the following were determined:

1. Area of nesting habitat area;

2. Features of habitat patch structure :

3. Boundaries frequently used by wolves (rivers, collectors, motorways and bridges, etc.). The estimated boundaries of habitat areas were plotted on the map with a dashed line. Boundaries of breeding areas were also marked, but with shorter dashes.

Some boundaries of family and pack habitats were determined presumptively. With the help of the mapping schemes, it is possible to calculate the number of wolves.

Thus, for successful mapping of wolf habitats, control and regulation of its numbers, it is necessary to accumulate a sufficiently large number of observations and actively use the obtained information. Qualitative analysis of the collected material allowed us to study in more depth the ecological peculiarities of the wolf's lifestyle in the Southern Priaralie. Having the materials of mapping, it is possible to carry out targeted conservation measures and rational regulation of wolf populations.

Field studies of wolf ecology were carried out according to the generally accepted method of Novikov G.A. (1953). The age of animals was determined according to the method of Klevezal G.A. (2007).

To determine the geographical variability of the wolf population, the methodology described in D.I. Bibikov et al. (1985).

When studying wolf nutrition, caprological analysis of excrement was carried out in the laboratory of the Karakalpak Republican Centre for Prevention of Quarantine and Especially Dangerous Infections of the Ministry of Health of the Republic of Uzbekistan. The population was also interviewed about wolf attacks on domestic animals and economic damage to farms in the studied districts of the Republic of Karakalpakstan.

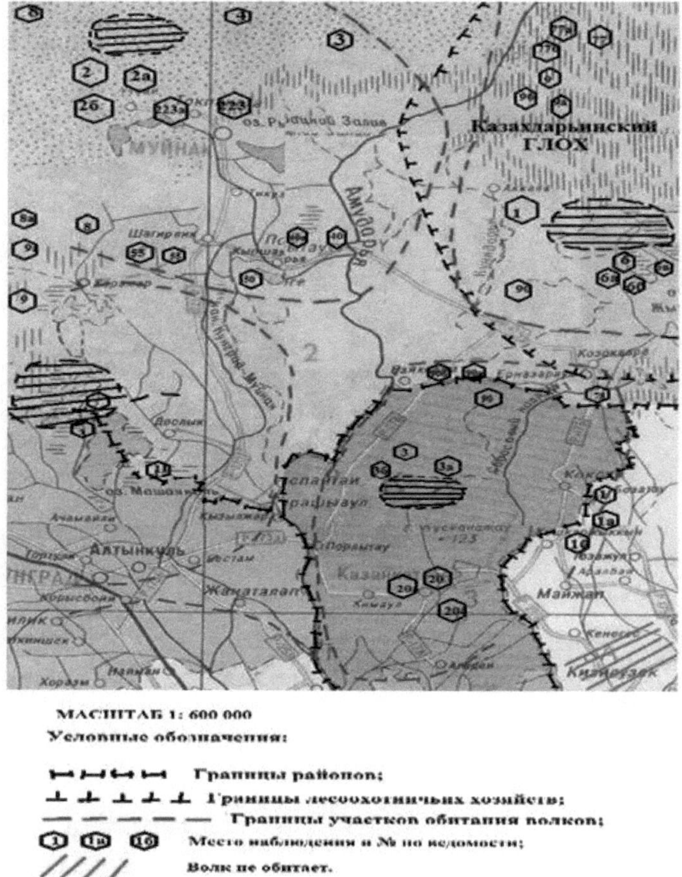

Fig.1. Map-scheme of wolf accounting in the Republic of Karakalpakstan

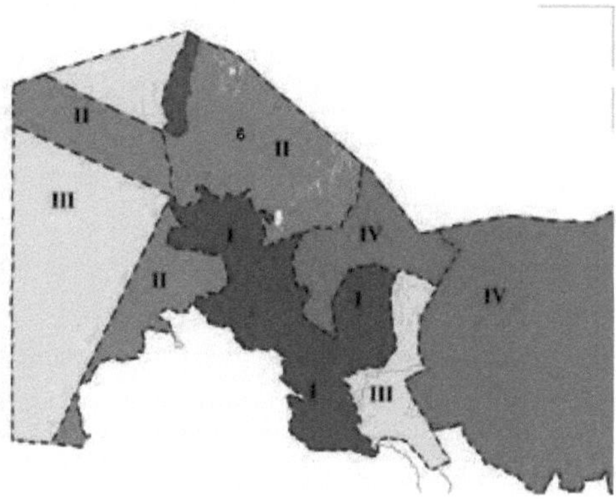

Fig.2. Map-scheme of zones of the Republic of Karakalpakstan in terms of quantity and quality of received information on wolves
I.   Area of continuous counting
II. Sufficient information zone
III. Area of insufficient information
V.   An area of almost complete lack of information

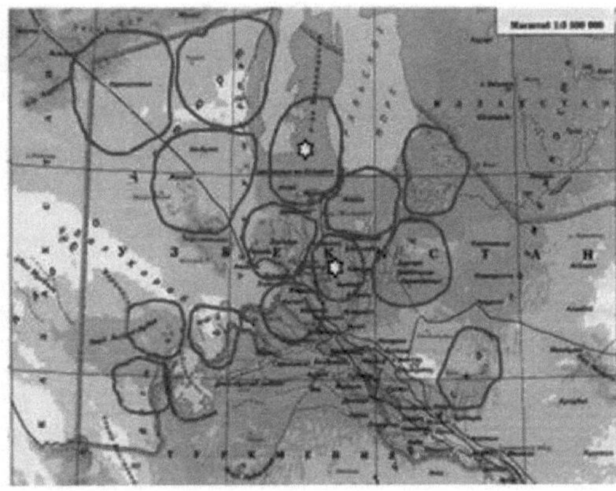

Fig. 3. The main indigenous (family and pack) areas of wolves on the territory of the Republic of Karakalpakstan

-Model plots of family wolf packs

- Site boundaries of individual family wolf packs

The dynamics of the wolf population dynamics in the study area was analysed according to the occurrence of wolves in the study area, by prey and pelt harvesting, as well as by causing damage to wild and domestic animals.

In the course of the work, an information network of observers was created from among the employees of the Khokimiyats of livestock departments in the studied districts of the Republic of Karakalpakstan, members of the Karakalpak branch of the Society of Hunters and Fishermen of Uzbekistan, employees of the State Committee on Ecology and Environmental Protection of the Republic of Karakalpakstan, the Inspectorate for Control over Protection and Use of Biodiversity and Protected Natural Areas under the State Committee on Ecology and Environmental Protection of the Republic of Uzbekistan, and an employee of the State Committee of the Republic of Uzbekistan. The obtained factual material was statistically processed according to G.F. Lakin (1980) and N.A. Plokhinsky (1970) using Microsoft Excel and STATGRAF computer programmes.

2.2 Physical and geographical characteristics of the Southern Priaralie
The studied region of the Aral Sea region and the adjacent vast territories represent a single complex with peculiar climatic, hydrological, soil and botanical conditions.
features.

The Priaralie refers to the landscape complex covering the north-western part of the Kyzylkum, the Zaunguz Karakum (including the Sarykamysh depression), Southern Ustyurt, and the Amu Darya valley and delta. The Aral Sea region covers an area of 245,500 km^2 , which is 19.2 per cent of the territory of Central Asia. It includes the whole Republic of Karakalpakstan, Khorezm province, Tashauz province of the Republic of Turkmenistan [99, p.96].   On physical-geographical , soil-climatic, geobotanical features within the studied region of the Aral Sea region can be distinguished three zoogeographical districts: Lower Amudarya, South Ustyurt and Kyzylkum. Each district is characterised by diversity of geological structure and landscape, climate and hydroregime, flora and fauna, as well as different intensity of anthropogenic factor. In general, they have similar features of structure and dynamics, history of formation, specific conditions of the natural environment.

The Lower Amudarya District borders on the Ustyurt Plateau in the west, Kyzylkum in the east, Zanguz Karakum in the south, and the dried bed of the Aral Sea in the north. It is characterised by flat relief and is entirely within the desert zone. According to the nature of combination of landscapes in hydrological and soil-botanical respect it is divided into southern (Khorezm-Turtkul), northern (Chimbay-Turtkul) and Primorsky parts [99, p.96].

The Kyzylkum district is represented by a sandy desert, predominantly flat in character, with small uplands of Tertiary-Cretaceous age. This district includes Sultanuizdag, Beltau and North Kyzylkum areas.

South Ustyurt district is located in the extreme north-west of the Republic of Karakalpakstan. According to climatic and soil-geobotanical conditions it is divided into Assakeaudan, Kosbulak and Churuk districts.

Lower reaches and Amu Darya delta

The lower reaches and delta of the Amu Darya is a huge alluvial plain located in the lower reaches of the river from the Tuyamuyun gorge to the Aral Sea. In the north it borders with the dried part of the Aral Sea, in the west it is bounded by the Ustyurt Chink, in the south by the Zanuguz Karakum and Sarykamysh Basin, and in the east by the Kyzylkum Desert.

The lower reaches of the Amu Darya differ significantly from the surrounding sandy and gypsum deserts in terms of physical and geographical conditions. According to researchers' data, the vast area of the Amudarya valley and delta is formed by alluvial-lake sediments, 365 km long, up to 319 km wide (up to the Sarykamysh basin), total area 45200 km^2 [99, p.96].

Geomorphologically, this territory is an ancient Amu Darya delta, which at Cape Takhiatash passes into the modern Priaralie lowland. The flat relief is sometimes broken by rare, far apart residual uplands: Kubetau, Jumurtau, Porlytau, Kushkanatau, Beltau, as well as the ancient beds of the Daryalyk and Daudan.

Under the influence of centuries-long human economic activity, the natural weakly undulating relief of the lower reaches of the Amudarya River has turned into a shallow terraced surface crossed in various directions by wide main canals and collector-drainage network.

Soils of the Amudarya valley are desert-takyr, sandy and desert-meadow, developed on ancient alluvium, and in the deltas are swampy, swamp-meadow and solonchak soils, and on the dried Aral Sea bed are solonchak soils, sandy and clay soils are found on uplands and in Tertiary remains. Small areas are occupied by grey-brown soils.

The lower reaches of the Amu Darya River are divided into two parts: southern and northern. The southern part includes Khorezm province, Turtkul, Ellikala, Beruni and Amudarya districts of the Republic of Karakalpakstan and Tashauz province of Turkmenistan. The northern part includes seaside territories, which represent a peculiar biocomplex (Amu Darya estuaries, local water bodies, Karabaili archepalag, etc.). The southern and northern parts of the Amu Darya lower reaches differ from each other in relief, hydrological regime, soil and vegetation cover.

The southern part (Khorezm oasis) of the lower Amu Darya is located on the western edge of the Kyzylkum between the Ustyurt Plateau, the Sultanuizdag Range and the Karakum Mountains. Geomorphologically, the region is an ancient delta of the Amu Darya. The formation of meso-microrelief is closely connected with the activity of the river and centuries-old culture of agriculture [118, p.202-209; 99, p.96].

The absolute height of the oasis above sea level in the south-east is 100 m, in the north-west -75 m. Karakum and Kyzylkum rise above the valley for 525 m, transition to the plain is usually smooth, in some places sands cut into the valley. Groundwater level is determined by collector-drainage and irrigation network, atmospheric precipitation has no noticeable influence. Groundwater and discharge water accumulate in numerous shallow depressions and broad depressions, resulting in the formation of a network of saline and shallow lakes, especially in the south-western peripheral part of the delta adjacent to the Karakums. Dry hot summers, severe cold, snowy winters, and high evapotranspiration are characteristic features of the climate of this oasis.

According to long-term data, the average January temperature is 4-5° C, which is 2.5-3.5° C higher than in the northern regions, winter is shorter by almost a month, absolute minimum -26° C, summer is hot, absolute maximum reaches up to 42.6^0 C (Fig. 4).

In terms of precipitation, these areas are among the driest areas of Central Asia. The average annual precipitation is about 80 mm, mainly in the form of rain in spring and autumn. Snow cover is extremely unstable and in some winter months lasts less than 10 days.

The duration of the frost-free period averages 214 days. Relative air humidity, due to the proximity of the Aral Sea, is higher than in the upper and middle reaches of the Amu Darya in the summer months - about 30-40% (Figure 5).

Figure 4. Dynamics of average annual air temperature in the southern part of the Amudarya Downstream (2009-2016)

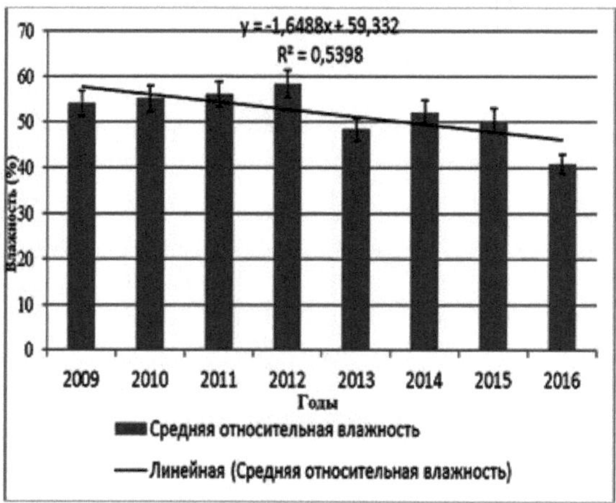

Fig.5. Dynamics of average annual relative air humidity (%) on the territory of the Southern part of the Lower Amudarya River (2009-2016).

Irrigation and soil cultivation drastically change water regime and determine the character of vegetation, create peculiar microclimate. The basis of vegetation cover of cultivated areas of oases and lower reaches of the Amu Darya River is created by agricultural crops (cotton, rice, corn, alfalfa, horticultural crops, etc.), gardens and forest and park plantations, and of non-tilled areas - by types of wild vegetation close to natural formations. In addition to weeds, the following vegetation groups are found: riparian, solonchak, psammophilous-shrub, halophytes and hydrophytes (reed, hornwort, sedge).

The southern part of the lower reaches of the Amu Darya is the most developed and populated, so natural vegetation in oases does not occupy large areas. Formations of camel's thorn and licorice develop on the deposits that have temporarily left cultural use.

Solanaceous formations are developed on downy solonchaks, margins of saline lakes: sarsazan, soleros, tamarix and annual solanaceae, cereals.

The floristic composition is relatively poor, with weeds occupying the leading position. Takyr and solanaceous vegetation is most widespread. Changes in the vegetation cover of the lower reaches of the Amu Darya in the ancient delta and oasis are partly related to the process of drying out of the territory after the cessation of irrigation and the gradual transition of perelogs into fallow lands, which determines the sequence of change of formations: yantachnaya, itsigekov, keireukovaya, biurgunovaya. On the sands and residual uplands there are wormwood-solanaceous and psammophyllous-shrub formations consisting of turangil, dzhida, tamarix, and chingil. They occupy a large area along the Amu Darya, large irrigation canals and collectors.

The vegetation cover of the Zaunguz Karakum adjacent to the oasis consists mainly of xerophytic formation (saxaul, cherkez, kandym, sand acacia, ephedra, wormwood, astragalus and some solanaceae, ephemera and ephemeroids). Herbaceous ajrek associations develop on non-wetlands and deposits. Reed, hornwort, various sedges, reedgrass, comb and other common components of herbaceous tugai grow along the lakeshores.

The animal world is poor, mammals are rare: wolf, jackal, fox, steppe cat, etc. There are many waterfowl, but most of them are migratory. There is a lot of waterfowl, but mostly migratory.

The northern (coastal) part of the lower reaches of the Amudarya River covers the flat territory of the modern delta from the Takhiatash district to the border of the dried bed of the Aral Sea. In the west it borders with the Ustyurt plateau chink, in the north it borders with the dried bed of the Aral Sea, in the east it borders with Kyzylkum almost along the line of Beltau and Sultanuzdag uplands.

The relief is a slightly convex plain with an extensive background of separate islands, some of which are remnant elevations of the Tertiary and Cretaceous periods: Krantau, Porlytau, Kushkanatau, Kyzyl-Dzhar, Beltau. This territory was formed by the activity of the Amu Darya during millennia as a result of deposition of sand and silt sediments. The climate is dry, continental, with hot, rainless summers and humid winter and spring periods. Positive average monthly air temperatures prevail over negative ones. Absolute maximum air temperature is 40^0 C, absolute minimum is 25^0 C, sometimes down to -32^0 C. The frost-free period is 193-195 days.

According to our collected data, the average annual air temperature ranges from 10.3^0 C to 14.2^0 C. The highest value was recorded in 2010 and 2016. The minimum indicator was noted in 2014 (Fig. 6). The trend line indicates the stability of this indicator.

Precipitation in the form of rain and snow is low, not more than 100 mm. Snow cover reaches 10-15 cm thick, does not last long, and quickly melts. There are 20-30 days with snow cover per year.

Figure 6. Dynamics of average annual air temperature in the territory of the northern part of the Lower Amudarya River (2009-2016)

Multiyear dynamics of precipitation shows a gradual increase in recent years (Fig. 7). The maximum precipitation was recorded in 2012, 2015 and 2016. The minimum level was recorded in 2010 and 2011. The linear trend indicates a sharp increase in this indicator.

The Amu Darya coastal delta, according to Lopatin V.G., Denigina R.S. and Egorov V.V. G., Denigina R.S. and Egorov V.V. (1958), in the 50s consisted of 4 parts: southern, relatively remote from the sea (land area 697 km^2 or 69%); eastern (land area 937 km^2); western (923 km^2); central (335 km^2). In recent years, the area of the modern ("living") delta has changed dramatically. The regulation of the Amu Darya and Syr Darya runoff, intensive expansion of irrigated lands for cotton and rice, and reduction of annual runoffs of these rivers have further worsened the hydrological regime of the Aral Sea and river deltas and led to drying up of the Aral Sea and the Amu Darya delta.

Fig. 7. Dynamics of average annual precipitation (mm)
on the territory of the northern part of the Lower Amu Darya Basin
(2009-2016).

In the 50s, the Amu Darya delta occupied 45000-50000 km^2 , of which about 35000 km^2 were water bodies fed through natural Amu Darya channels [106, p.10-17; 90, p.35-53]. In 1963, the area of open water surface (clear water mirror, or pleats) of delta water bodies was 98 thousand ha [120, p.17]. Subsequent reduction of the Amu Darya flow and lowering of the Aral Sea level resulted in drying up of more than 40 lakes with a total area of 50 thousand ha [5, p.80].

At present, all water bodies of the Karabaily archipelago, the Bozatau and Karadjar lake systems are almost dried up. Due to the drying up of the Aral Sea, the Muynak, Sarbas and Abassi bays have dried up in some places. The beautiful muskrat lands in the lower reaches of the Amu Darya, where more than 1 million muskrat skins were harvested annually in 1956-1957, have completely dried up. Since 1977-1978, due to the small number of muskrat skins, they have not been harvested.

Deterioration of the hydrological regime and land development has led not only to the reduction of areas occupied by riparian forests and reed beds, but also to the replacement of hydrophytes by xerophytes and halophytes. Reed, hornwort, false reedgrass and water-logged species grow in flooded areas. On the banks of the Amu Darya River, turangil and sucker associations, shrubs and semi-shrubs prevail in the floodplains. This creates favourable conditions for a variety of animals, including mammals. Wild boar, badger, fox, jackal, wolf, reed cat, hare, weasel, and steppe polecat are often found here, and waterfowl, both nesting and migratory, are especially abundant.

Ustyurt Plateau

Ustyurt occupies about 200 thousand km^2 , located between the Aral Sea and the Caspian Sea. It is located 50-280 metres above sea level. The plateau is bordered by high cliffs, or chinks. Its surface is composed of Sarmatian sediments. The Karakalpak part of Ustyurt is 7 thousand km^2 . It is a flat vast plain cut by uvalas and depressions, the largest of which - the Assekeaudan and Barsakelmes depressions - lie below sea level. Steep chinks (90-100 m, sometimes 240 m of absolute height) break off the plateau towards the Aral Sea. From Cape Urga, the chink runs southwards along the western shore of the former Aibugir Basin. On the southern side of the chink are the Sarykamysh depression, the Kaplankyr upland and the ancient bed of the Amu Darya Uzboy [99; p.96].

According to the physiographic zoning, the Ustyurt Plateau is considered as part of the Central Kazakhstan Province and is distinguished as an integral physiographic unit by the unity of geological-geomorphological and paleogeographic conditions of formation of modern landscapes [29, p.106-111].

Landscape   ,   climatic,   soil and geobotanical   .

Ustyurt conditions are divided into three districts (Figure 8.). The North-Ustyurt district covers the northern part of Karakalpak Ustyurt. It is a gently undulating plain with heights up to 150 metres. The predominant type is the landscape with biyurgunovo-boyalyshev complex, which occupies up to 90% of the whole territory of the district. The Central Ustyurt district occupies the central, most depressed part of the territory. Absolute elevations of the territory vary from 71 metres at the bottom of the Barsakelmes Basin to 150 metres in the north. In the north-west, the Barsakelmes depression continues in the form of a lake-accumulative plain composed of sandy sediments. The South Ustyurt district occupies the territory south of the Karabaur uval and the adjacent upland. The relief of the area is undulating plain, dissected by runoff troughs. In the southern part of Ustyurt there is a large Assakeaudan depression, whose precipitous slopes reach a height of 40-50 metres. The vast bottom of the depression is characterised by gently undulating relief. Here, as well as in depressions adjacent to Barsakelmes, semi-fixed knobby sands 2-5 m high made of fine-grained sand particles and salt dust deposited around shrubs are widespread [29, p.106-111].

Air temperature is one of the important indicators in characterising the landscape.

According to our research on the territory of the Ustyurt Plateau for the period from 2009 to 2016, the air temperature change was analysed (Fig. 9). The average annual air temperature is quite stable for the entire study period. Note that the maximum temperature was recorded in 2014 -35.1^0 C, and the minimum temperature reached -2.9^0 C in 2009 and 2014.

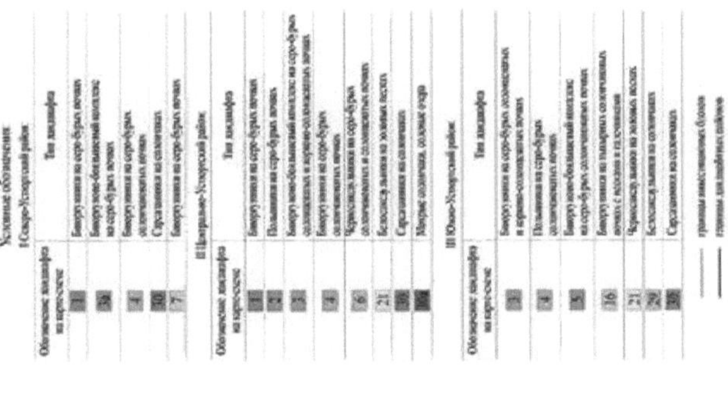

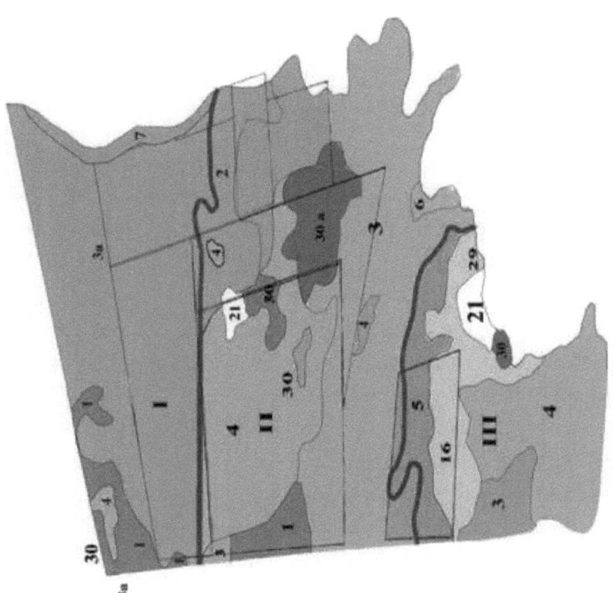

Figure 8. Landscape map (by Kleimenova, 2010)

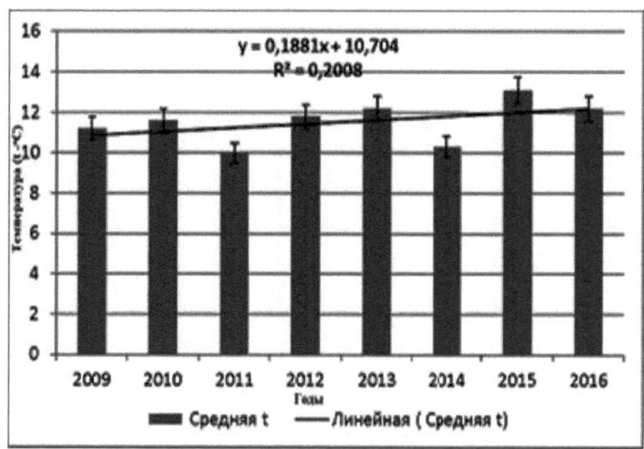

Fig. 9. Dynamics of average annual air temperature in the territory
Ustyurt Plateau (2009-2016).

By analysing the annual average relative humidity, it can be noted that the maximum indicator is in 2009 (57.3%), in the rest of the period there is a steady decrease until 2014 (43.8%) and then there is a gradual increase to 48.4% in 2015 (Figure 10).

The amount of precipitation of any landscape is an integral part due to its impact on flora and fauna, as well as play a major role in the process of water exchange cycle in the environment. The analysis of precipitation indicators of the Ustyurt Plateau for the period from 2009 to 2016 showed that two maximum peaks of precipitation -2012, 2014, 2015 (up to 35.1 mm) were identified (Fig.11).

According to soil-geographical zoning, the territory of Karakalpak Ustyurt belongs to the sub-boreal belt, the zone of deserts and semi-deserts. The following soil varieties are represented on the territory: grey-brown desert, takyr, takyr, gypsozems,

desert sandy, desert solonts, solonchaks. Salinity and low humus content are characteristic features of soils of the Ustyurt plateau [29, p.106-1111].

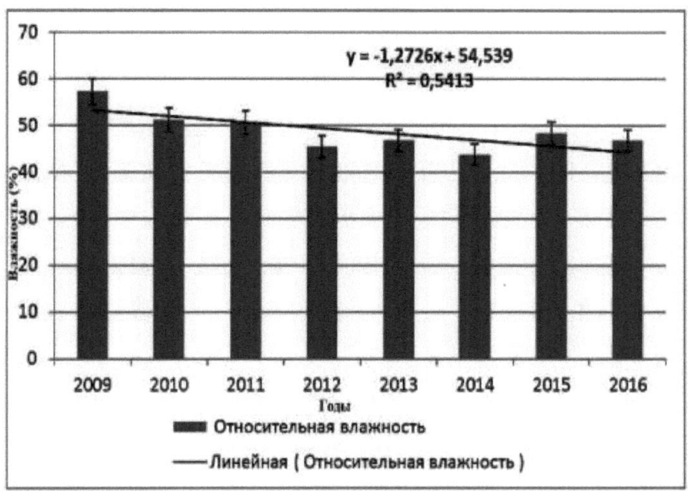

Fig. 10. Dynamics of average annual relative air humidity (%) on the territory of the Ustyurt Plateau (2009-2016).

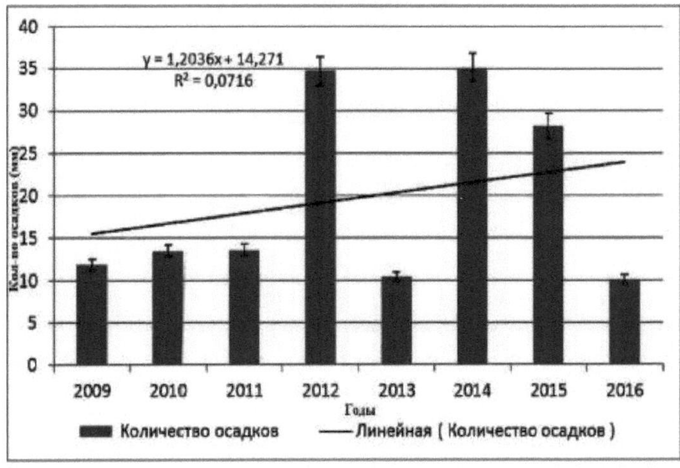

Fig. 11. Dynamics of average annual precipitation (mm) on the territory of the Ustyurt Plateau (2009-2016).

According to the data of specialists, ecological zoning of soils of the Ustyurt plateau was carried out [29. p.106-111]. Areas with very low soil stability are mainly confined to the Barsakelmes Shor in the Central Ustyurt landscape region. In the rest of the considered territory they are represented to a limited extent, within the Assakeaudan depression of the South Ustyurt landscape region and in the extreme north-west of Karakalpak Ustyurt. This category of lands includes solonchaks and crust solonts and their variations. Areas with low level of stability are represented mainly by takyr and takyr-like soils. Their low humus content, unfavourable physical and chemical properties, low intensity of soil processes and incompleteness of soil processes under contrasting natural and climatic conditions worsen their environmental sustainability.

Considering the dynamics of average annual soil surface temperature indicators on the territory of Ustyurt Plateau for the period from 2009 to 2016, it can be noted that the maximum indicators were registered in 2012 and 2014 (up to 43.3-44.2^0 C), and among the minimum indicators the lowest temperature was registered in 2009 and 2015 (up to - 5.6^0 C) (Fig. 12).

A significant array of moderately stable soils was formed in the north-western part of the Central-Ustyurt landscape area. The character of plant communities is determined by natural-climatic factors: intensive solar insolation and high summer temperatures, strong frosts combined with weak snow cover in winter, moisture deficit in soil and air in summer and sharply expressed daily and annual temperature fluctuations [84, p.200; 17, p.177]. According to scientists, the floristic composition of Ustyurt is up to 326 species of wild higher plants belonging to 192 genera, 42 families. According to the latest data, 183 species of higher plants belonging to 114 genera and 25 families have been identified [1, p.78-90]. In the Karakalpak part of Ustyurt there are plant species rare in Central Asia in general, known so far only in a few places, and *Salsola chivensis* is "Ustyurt endemic" [84, p.200]. [84, c.200].

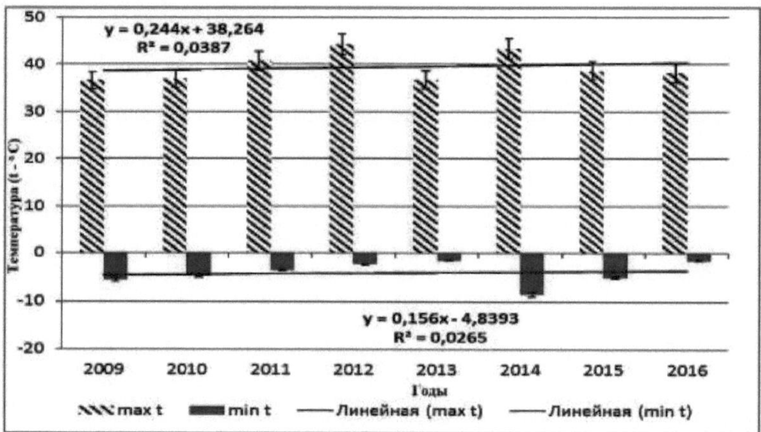

Fig.12. Dynamics of average annual soil surface temperature (max and min) on the territory of the Ustyurt Plateau (2009-2016).

The modern fauna of the Ustyurt Plateau includes 2 species of amphibians, about 35 species of reptiles, 200 species of birds, 44 species of mammals.

The data of experts indicate the habitats of rare plant species and the main habitats of protected species of animals and birds, areas of the state landscape reserve "Saigachiy", ornithological reserve "Sudochie", which is one of the largest transcontinental Asian-European migratory bird migration routes along the Eastern Chink of Ustyurt. Recent studies by ornithologists show that there is a large concentration of the relict Flamingo bird in this area.

North-western part of the Kyzylkum desert

The north-western Kyzylkum is a vast flat plain (75 - 100 m above sea level) occupying 350,000 km^2 of area within the Republic of Karakalpakstan, Uzbekistan and Kazakhstan. Kyzylkum is located between the Amu Darya and the Syr Darya. It is similar to the Zaunguz Karakum in terms of the history of formation of the landscape character, relief, peculiarities of flora and fauna.

The territory of the Republic of Karakalpakstan includes its north-western part with an area of about 2021 km^2 [99, p.96]. In the west Kyzylkum adjoins irrigated lands, in the extreme south-west borders with the Sultanuzdag range, in the north-west borders with the boundaries of the dried bed of the Aral Sea. The relief is diverse in form and genesis. Northern areas of the Karakalpak part of Kyzylkum are covered with fixed hilly-ridgy sands and wide dry riverbeds, and the strip adjacent to the delta with barchan sands. In the south rise low remnant massifs, the highest of them, Sultanuzdag, 485 metres high. Between them lies a flat plateau covered with ridge and ridge-bumpy sands. Sometimes there are extensive hollows, clay takyr surfaces of light grey colour, where there is no vegetation. Ridge sands are up to 5-10 m high and up to 1000 m wide. Humpy sands 1-5 m high, more than 12 m in diameter are formed by loose sands around shrub vegetation.

The formation of Kyzylkum relief was greatly influenced by weathering, precipitation and other climatic factors. The relief, in turn, determines the distribution of water, diversity of flora and fauna. The desert climate is sharply continental with dry hot summers and relatively cold snowless winters. The average amount of precipitation throughout the territory of Karakalpakstan is about 100-110 mm per year, falling mainly in the winter-spring period. The average annual temperature is 12 degrees Celsius (maximum plus 41.5 - 45, minimum minus 22-28). Fig.13 shows the average annual dynamics of air temperature in the territory of the north-western part of Kyzylkum.

Fig. 13. Dynamics of average annual air temperature in the territory of the north-western part of Kyzylkum (2009-2016).

The dynamics shows the maximum values of average air temperature in 2013, 2015,

2016. The minimum value was recorded in 2014. The linear trend indicates a stable temperature level. Drought is a scourge not only for Kyzylkum, the duration of the hot period here reaches from 200-215 days to 250-260 days.

In connection with the drying up and disappearance of the Aral Sea, desertification has become one of the main problems, as a negative phenomenon that removes huge suitable land resources from the productive state. Increased drought caused by the drying up of the Aral Sea has led to changes in the landscape, deterioration of environmental conditions, the probability of occurrence of which is 70-80% in the territory of Kyzylkum. The analysis of precipitation indicators for the territory of the north-western part of Kyzylkum for the period from 2009 to 2016 showed that one maximum peak of precipitation -2016 (up to 15.4 mm) was identified (Fig.14). The linear trend shows a gradual increase in precipitation in this area.

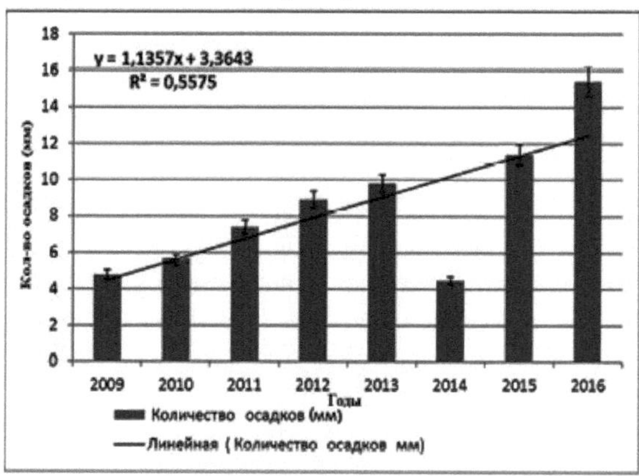

Fig. 14. Dynamics of average annual precipitation (mm) on the
territory of the north-western part of Kyzylkum (2009-2016).

Fig.15 shows the dynamics of average annual wind speed on the territory of the north-western part of Kyzylkum, which indicates the great ecological importance of this factor. Analysis of dynamics showed a gradual slow increase in wind speed for the considered period. The linear trend also indicates an increase in this indicator.

Soil cover of the north-western part of Kyzylkum is represented mainly by grey-brown soils. Vegetation cover consists of about 900 species of higher plants. Of these, the complex of saxaulo - rank association is characteristic. The largest areas in Kyzylkum are occupied by shrub and semi-shrub vegetation and wormwood formation.

In sandy deserts psamophilous woody-shrub thickets consisting of black and white saxaul, cherkez, dzhuzgun, sand acacia, etc. are widespread. Artemisia, boyalych, teresken, and some species of saltwort are widespread.

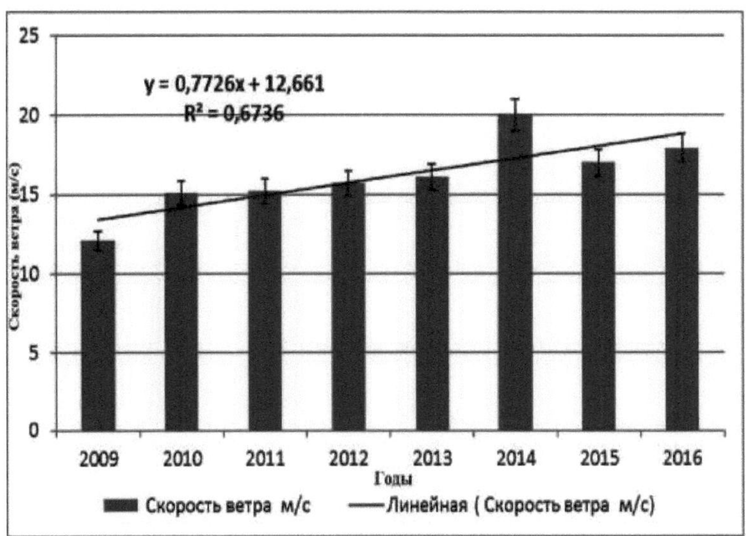

Figure 15. Dynamics of average annual wind speed (m/s) in the territory of the North-Western part of Kyzylkum (2009-2016).

In order to divert collector-drainage water from the southern right-bank districts of Karakalpakstan towards the Aral Sea, drainage and irrigation infrastructure was reconstructed to restore wetlands. The total length of collector canals within Kyzylkums exceeded 250 km. Annual volumes of diverted collector-drainage water depend on water availability, in high-water years they exceeded 750 million cubic metres. Many wetlands were formed along the collector route, their total area exceeded 9 thousand hectares. Watering of the North-Kyzylkum region led to radical transformations of its ecosystems: waterlogging occurred, reed and cattail thickets appeared on the banks of wetlands, rare tamarisk and akbash bushes prevail in the coastal area.

The fauna of the northern part of Kyzylkum is represented by typical desert species. Here, psammophilous species predominate in the theriofauna. Of the 67 mammal species distributed throughout the territory of Karakalpakstan, 41 (59.4%) are found in large fixed barchans (insectivores -3; man-eaters -7; hares - 1; rodents -18; predators - 9; ungulates -3). Among birds there are desert warbler, common saxaul jay, desert finch, saxaul sparrow. Reptiles are numerous. Among predatory mammals jackal, fox and desert wolf are often found.

In sandy massifs there are 38 species of mammals, on the alluvial plain with dense soil there are 31 species of mammals, of which 18 are rodents (gophers, sandflies, gerbils, muskrats, etc.).

Thus, intensive desertification of Kyzylkum and desiccation of the Aral Sea led to ecological disaster of natural environment and living conditions of the whole Aral Sea region. But in recent years there are processes of improvement of natural environment due to measures on reconstruction of drainage and irrigation infrastructure with the

purpose of restoration and functioning of wetlands as important ecological factors for
sustainability of ecosystems in the Aral Sea region.

CHAPTER III

ECOLOGY OF WOLVES (*CANIS LUPUS* LINNAEUS) IN THE SOUTHERN ARAL SEA REGION

3.1. Systematics and distribution range

According to zoological nomenclature, the wolf belongs to the predator group *(Carnivora),* the dog family *(Canidae)*, the genus wolf *(Canis)*, and its Latin name - *Canis lupus* - was given to this predator by the famous Swedish systematist of animals and plants Carl Linnaeus in 1758. At present the question about taxonomic structure of genus *Canis* remains rather debatable. Including intraspecific groupings within the genus *Canis* proper are also not unambiguously interpreted.

The more widespread view among specialists is that all taxa with modern five species of *Canis (lupus, latrans, familiaris, rufus, hodofilax)* are included in one nominative subgenus *Caniss. str.* [8, c.605]. Different authors' ideas about the taxonomic boundaries of the "wolf group" are quite similar in the sense that species occupying deliberately far from the central position of *Canis lupus* were not included. The intraspecific system of *C. lupus* has not yet been satisfactorily developed. It is presented in full in the works of only two authors, Pocock (1935) and Mech (1974). In their works very limited data on geographical variability and species system of the wolf are collected. At the same time, they do not reflect materials on actual intraspecific relationships, which requires further elaboration. They defined 32 subspecies of the wolf, which hardly correspond to the true taxonomic structure of this species. The intraspecific systematics of the wolf was also considered in various regional summaries [86, p.293; 18, p.123-193].

For the last 55-60 years in the works of American scientists 20 subspecies of wolf are accepted according to the system proposed by Goldman in 1944. On the territory of Europe up to 12-15 subspecies are allocated, and on the territory of CIS a fractional system including 9 subspecies is allocated, and for Palearctic according to Geptner (1967) 6 subspecies of wolf are allocated. According to Ellerman and Morison Scott (1966) 12 subspecies are indicated for the whole Palaearctic.

Four subspecies are indicated on the territory of the former Soviet Union [86; p.293]. A more precise and generalising point of view is given by Corbet (1978), who believes that due to the clinal nature of variability of most morphophysiological characters in the Palaearctic, discrete subspecies cannot be distinguished. The construction of the intraspecific system of the wolf involves many complexities of a clinal nature. These include variability of body and skull dimensions, variability of hair colouration within the territory of distribution. Very great difficulties are associated with the scarcity of morphological material from different parts of the range (especially Siberia and Central Asia) and from areas where wolves have been completely or almost completely exterminated. Single cases were used in collecting and analysing the materials, which represented an impossibility to verify representative samples. Based on the results of the study by Bibikov D.I. et al. (1985) an intraspecies system of the wolf was precisely developed, in which all characteristics were taken into account that could, in case of insufficiency of morphological material and incomparability of diagnoses of separate subspecies, clarify or determine the general picture of subspecies differentiation. In his opinion, within the entire range, the diagnostic characters may be the candylobasal length of the skull, which can be considered as an indicator of its overall size, and the type of hair colouration.

Territorial changes in the size and colouration of wolves are very closely related to landscape and climatic conditions, which are adaptive in nature and reflect the specificity of adaptive variability of its intraspecific forms. The boundaries of subspecies should most often be combined with the boundaries of natural-geographical areas [8, p.605].

Based on the definition of the taxonomic structure of *Canis lupus* species according to Bibikov et al. (1985), in the territory of the Southern Priaralie there is the desert wolf *Canislupus. Desertorum* Bogdanov, 1882. In his opinion, the geographical variability of the wolf presents considerable fluctuations. According to Palvanazov (1974) the desert wolf *(C. l. Desertorum* Bogdanov, 1882)* the body length of adult males (9 individuals) captured in different areas of Central Asian deserts is 95 -116.5 cm, females (7 individuals) -92,8 -110,7 cm, tail length of males 29 -34,8 cm, females - 28,2 -34,0 cm, hind foot of males -17 -22,6 cm, females -16,5 -21,7 cm, ear height of males -7,1 -9,1 cm, females -6,7 -9,0 cm. Weight of males 19,5 -31kg, females -15,7 - 26,8kg. According to his measurements (in mm), the candilobasal skull length of males (9 individuals) 186 - 208.2, females (5 individuals) -184.1 -197.5, the total length of males 192.5 - 235.0, females -187.8 -219.9, the basic length of males 167,1 - 196,0, and females -172,8 -185, length of facial part of males 121 - 148, females -120 - 135, zygomatic width of males 95,3 -126,2, females - 92,1 -113,7, interorbital width of males 28 -41, females -26,1 -37,4.

The size of the skull of the desert wolf inhabiting the deserts of Central Asia is noticeably smaller than that of animals from Kazakhstan and Siberia (the candylobasal length of the skull of males is 15 - 20 mm shorter, that of females - 20 - 30 mm shorter). According to Geptner et al. (1967) the desert wolf, in comparison with other subspecies (tundra wolf, Siberian wolf) is characterised by smaller size and weight by 15 - 25 kg. The smaller size of this predator can be explained by the fact that it mainly hunts small ungulates (gazelle, saiga), hares and rodents.

The desert wolf has well expressed sexual dimorphism: males are larger than females. Its hair coat is coarse and short. The colouration is light grey with a sandy tinge, black endings of the spine hairs mainly along the ridge. A reddish tinge is clearly visible in the scutellum. The ventral side is lighter than the back. The forehead is light grey, with a faint sandy patina. The back of the head and the outside of the ears have a reddish tinge. The tail is covered with stiff hairs. The colouring of summer fur is somewhat different from winter fur. Thus, the pelts of wolves harvested in Ustyurt and Kyzylkum in July and August were dominated by a mixture of rusty and ochre tones, while those of animals harvested from December to February were more grey.

Within the modern territory of the CIS and neighbouring countries, about 1000 years ago there were three large, relatively isolated regions of wolf population: tundra, European-steppe (forest-steppe) and Central Asian. In each region, wolves were associated with certain groups of ungulates hunted: tundra - with reindeer, partly with snow sheep, forest-steppe - with ungulates of this landscape, Central Asian - mainly with semi-desert, desert and high-mountain ungulates (saiga, gazelle, etc.).

In the CIS, the desert wolf is distributed in the plains of Central Asia and Southern Kazakhstan. The northern border of its distribution reaches the middle reaches of the Emba River, Northern Priaralie, it inhabits Betpak-Dala, northern and southern Pribalkhashye, Alakul and Zaisan hollows [93, p.326, 94, p.76;124, p.95]. In the deserts of Central Asia predators were always less than in mountainous conditions. Their distribution here was limited by the lack of water sources and shelters. In addition, wolf distribution in the deserts of Central Asia is connected with pasture cattle breeding, abundance of wild ungulates, availability of convenient places for breeding young and drinking water. It is most widespread in Ustyurt, in the floodplain of the lower reaches of the Amu Darya, around small lakes on the Aral coast, in the west of Turkmenistan, in the foothills, where there are conditions for its sustainable existence. In winter, wolves move to the floodplains of rivers where cattle, wild boars, hares and other prey are concentrated, and in spring they follow flocks of sheep. In summer, the wolf is absent in the waterless areas of central and southern Kyzylkum, Zaunguz and eastern Karakum and in the Uchtagan sands. Outside the CIS, the habitats of the desert wolf are Afghanistan and Iran. Fossil remains of wolves have been found in different places of the desert. In the Quaternary sediments of the Zarafshan valley, fragments of the skull and lower jaw, teeth, etc. were collected during excavations of the Pejikent quarry. - A total of 115 fragments from nine individuals. The species composition of the fauna and the character of bone preservation allow us to attribute the Penjikent burial to the Middle Pleistocene [4, p.60-63].

Thus, the wolf has been present in the deserts of Central Asia since the Pleistocene. The wolf from the Penjikent quarry belongs to a small form close to the desert wolf.

During the study of the desert wolf, we established the distribution range and approximate density of this predator in the Southern Priaralie. The northern boundary of the distribution area includes the north-western part of Ustyurt (Karakalpakia, Jaslyk), Muynak (Uchsai, Kyzylzhar, Karadzhar, Shege, Kazakhdarya) and Kungrad districts. The southern border includes Turtkul, Beruni, Ellikala districts (Jambas kala). The central distribution area includes the territories of Takhtakupyr, Chimbay, Karauziak, Kegeyli districts. It should be noted that the territories most occupied by wolves are Kegeyli, Chimbay, Kungrda and Muynak districts. This is mainly due to the development of cattle breeding, stable food base (rodents, birds, reptiles, as well as plants). We have noted cases of steppe wolves from Kazakhstan and Turkmenistan entering the territory of Ustyurt and Dzhanbas kala in search of food and in pursuit of saigas (Fig. 16).

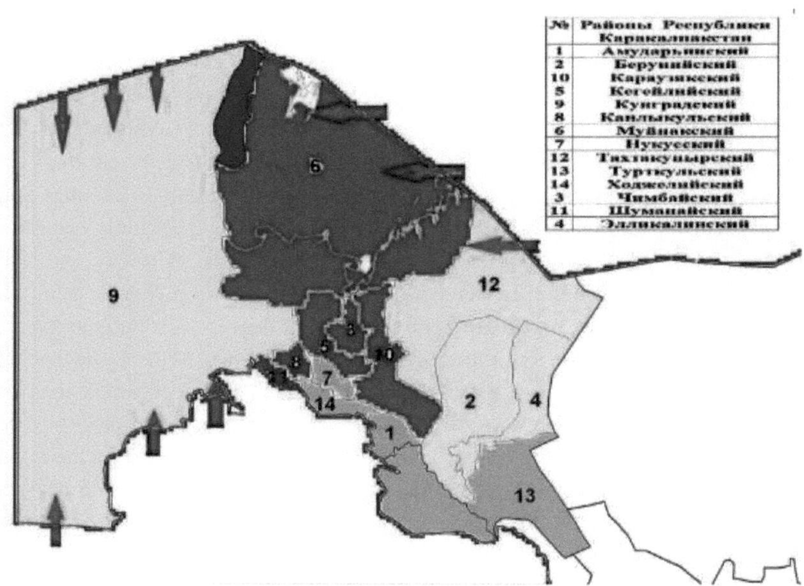

№	Районы Республики Каракалпакстан
1	Амударьинский
2	Берунийский
10	Караузякский
5	Кегейлийский
9	Кунградский
8	Канлыкульский
6	Муйнакский
7	Нукусский
12	Тахтакупырский
13	Турткульский
14	Ходжейлийский
3	Чимбайский
11	Шуманайский
4	Элликкалинский

Fig. 16. Range and density of wolf population in the Republic of Karakalpakstan

The estimated wolf population density is 0.05-0.06 individuals per 1000 ha.
The estimated wolf population density is more than 0.05- 0.06 individuals per 1000 ha.
The estimated wolf population density is less than 0.05-0.06 individuals per 1000 ha.

Entries of steppe and desert wolves from the territory of Kazakhstan and Turkmenistan
According to Palvanazov M. (1974) relatively high density of wolf population was observed in biotopes of gypsum desert (Ustyurt, Betpak-Dala and foothills bordering with deserts), on chinks and uplands, in reed thickets. In these biotopes wolves found optimal living conditions. Here a considerable number of wild and domestic ungulates, hares, rodents and birds were available for the predator; here they were also provided with water supply round the clock. Such biotopes represented reserves, from which the predator's reserves are replenished, as it is there that the greatest number of its brood holes is available. Recently, however, the character of the wolf distribution area has changed somewhat due to the intensive drying up of the Aral Sea and the lakes adjacent to it. These conditions have led to a reduction in the predator's range, and their concentration has been observed mainly near water sources, where there is a more acceptable food base.

In the course of expeditionary studies we found traces of wolf activity near the coast of the South Priaralie wetlands. According to our observations wolves often stay along the eastern chink of Ustyurt - near Sudochie, Karateren, Zhaltyrbas, Garkyldak kul and other lakes. In deserts, wolves prefer to stay in open spaces with low vegetation, where they can see prey or detect danger from afar. In these territories, wolves mainly stick to ravines, hilly sands overgrown with saxaul, but always close to water (Fig. 17).

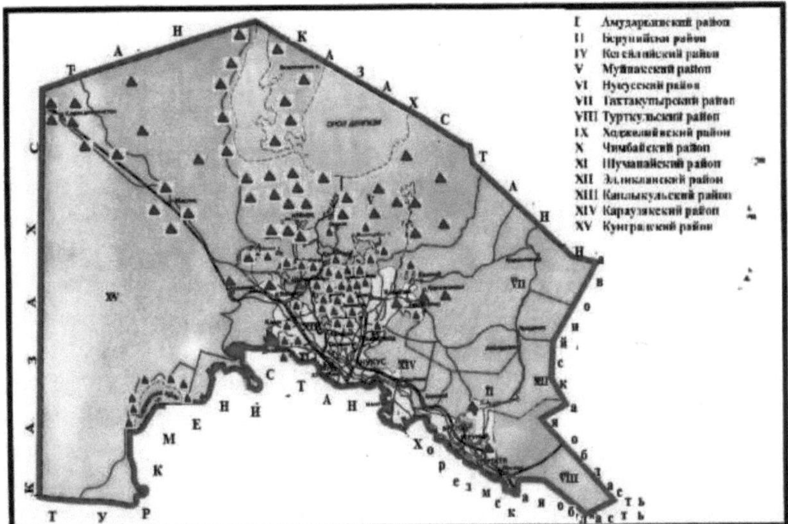

Figure 17: Wolf occurrence in the territory of the Republic of Karakalpakstan according to the results of the questionnaire and survey method A Wolves encountered According to questionnaire data and our observations in the desert zone, wolves often stay near artesian wells, where other species of animals (gazelles, saigas, etc.) come for watering, as well as near the camps of shepherds and flocks of cattle breeders.

Specialists note that in the Amu Darya floodplains, this predator's denning sites were mainly confined to bushes and reeds. It was often found near the shores of lakes, where it fed on fish remains and hunted waterfowl. The wolf avoided monotonous continuous riparian thickets, preferring places where small patches of stands alternated with glades [93, p.326].

According to our data, due to the fact that the territory of tugai thickets have sharply reduced and most of them are transferred to protected areas and forest hunting farms the frequency of wolf sightings has decreased. This is explained by the fact that anthropogenic disturbance factor is very important in their dispersal in these territories.

Thus, it can be noted that the distribution area of the wolf population in the Southern Priaralie is closely related to the presence and location of water bodies and optimal forage base. The size of the occupied territory of wolves varies depending on the availability of forage conditions.

3.2. Nutrition

Nutrition is one of the most important conditions in the life activity of mammals. The nature of nutrition is determined by the attitude of a given species to the sources of necessary food substances and determines the position of this animal in biocenoses -

natural groupings of organisms [3, p.396]. The forage ration, first of all, depends on the forage availability of the wolf habitat. It is established that the wolf diet includes both wild and domestic ungulates, including the hare - tolai, rodents, reptiles and birds. In addition, their diet includes predatory mammals (fox, corsak, cats, jackal, dogs). When food is scarce, it preys in small quantities on fish, insects, mollusks, and eats plants (suckers, vegetables, cereals). According to our data, the wolf labilely uses forage resources in all natural zones of its habitat.

During the study, the frequency of occurrence of a certain type of food was recorded, i.e. the qualitative composition of the diet was assessed. According to Palvanazov (1974), wild ungulates occupy on average 41.0 % of the wolf's annual diet in Ustyurt: saiga 23.7 % and gazelle 17.3 %. The main prey of the wolf in Ustyurt is saigas and small mammals such as tolai hare, yellow gopher, greater sand lance, noon vole and other rodents. He also noted that gazelles, due to low numbers, have practically dropped out of the wolf diet.

As a result of our analysis of the qualitative composition of the wolf forage base, it was found that in Ustyurt, the predominance of natural forage is 80%, including saiga and gazelle, and in some places wild boars, while forage resources of anthropogenic origin occupy about 20% (Fig. 18. A). A comparative analysis with the data of Palvanazov (1974) showed that there is a decrease in the consumption of wild ungulates due to a decline in their numbers. The second place after ungulates in the wolf's diet belongs to the hare-tolai.

In the Southern Priaralie, the hare-tolai abundance is of significant importance for the wolf, as according to specialists [93, p.320] it was noted that in years of increased hare abundance, the number and fatness of the wolf is higher than in years of low hare-tolai abundance.

During expedition trips during the years of the study we noted a large number of hares in the Central part of the Southern Priaralie. The wolf excrements we found (N 43° 15^{I} 46, E 059° 19^{I}) also contained hare hair remnants. This is in agreement with the data of Bibikov D.I. et al. (1985) on food specialisation, stating that the occurrence of hares in wolf faeces collected in Ustyurt is 12.4%, 16% in spring, 4.6% in summer and 16.4% in autumn. He also noted that wolves are less likely to pursue wild ungulates in years of hare abundance.

According to literature data [93, p.326, 94, p.76], wild ungulates noticeably lose their priority in the predator's diet as it moves to the south-eastern part of Karakalpakstan, while domestic animals and small mammals become the main components of the diet. As evidenced by the results of our studies, in the territory of the Southern Priaralie this trend is manifested not only in the direction from north to south, but also in the direction from west to east.

Quantitative analysis of the wolf diet in the Central part of the Southern Priaralie region showed that the main share belongs to forage resources of anthropogenic origin (about 56%), and the remaining share of 44% - to forages of natural origin (Fig. 18. A, B).

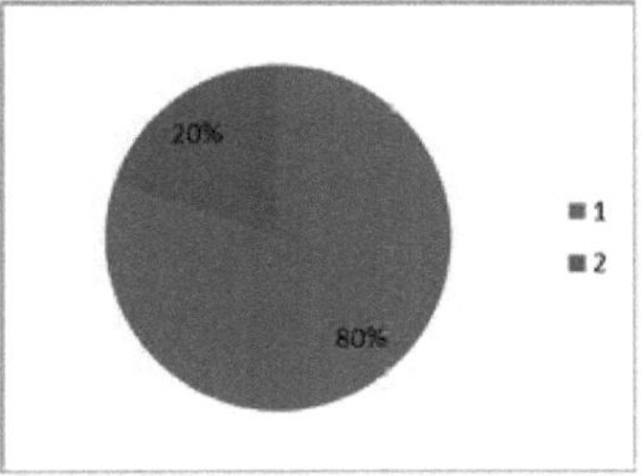
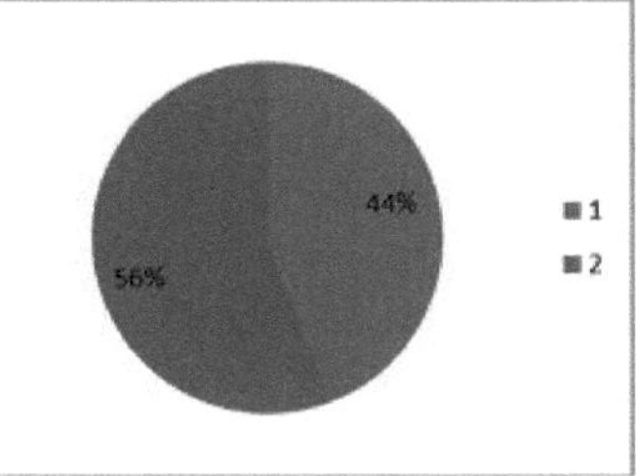

Fig. 18. Ratio (%) of natural (■ 1) and
anthropogenic fodder (■ 2) in the diet of wolf in different natural zones of the
Southern
Priaralie

Note: A-North zone, B-Central zone

The main difference in foraging specialisation of different
wolf populations
in the Southern Priaralie concerns the ratio of the frequency of
occurrence of anthropogenic and natural resource in the
The diet of the predator, while differences in the species composition of natural forages are of secondary importance (Fig. 19).

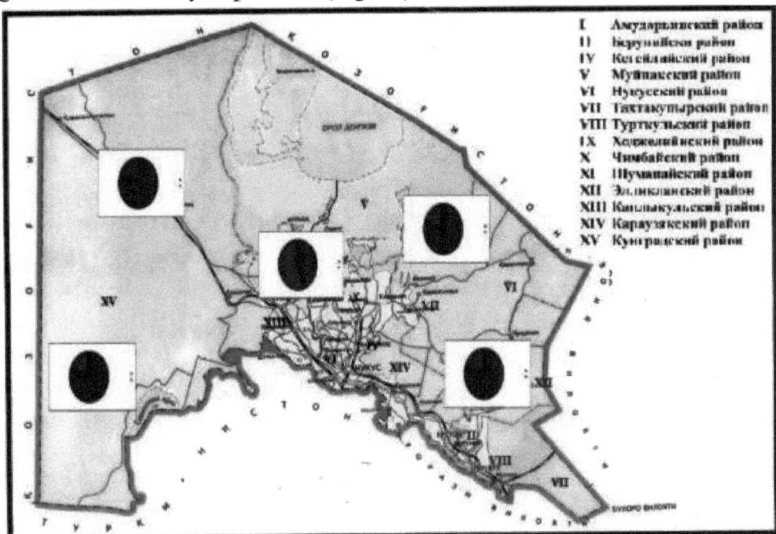

Fig.19. Ratio (%) of natural (□ 1) and anthropogenic forages (2) in the diet of wolf in
different natural zones of the Republic of Moldova
Karakalpakstan

According to Bibikov D.I. et al. (1985) the main domestic animal victims of the wolf
are the following descending series: sheep, calves, foals, donkeys, cows, horses, dogs

and poultry.

Certain differences in the feeding spectrum of packs and solitary individuals and temporary associations of pereyarka have also been established. In general, domestic ungulates dominate in the wolf diet in the Central part of the Southern Priaralie due to intensive development of cattle breeding. According to questionnaire survey data in the territory of Uchsay (Muynak district), the frequency of solitary wolves is high. Solitary individuals prefer domestic animals mainly small (dogs, cats, goats, sheep, foals) as easier prey in comparison with wild ungulates, hunting of which requires a partner. Temporary associations of pereyarka are characterised by a significant percentage of domestic animals in the diet.

During the expedition along the route Nukus - Turtkul - Janbas kala massif (about 240 km) from 2009 to 2016, we explored the territory of JSC "Kasym Nurumbetov", located in Turtkul district, in the south-eastern part of the Republic of Karakalpakstan. Although the territory is located in a desert zone, but due to small lakes it is characterised by features of an oasis with rich flora and fauna.

With the support and assistance of farmer Matirzaev Azamat, surveys were conducted among shepherds and residents of Dzhanbas kala settlement. The results of the survey revealed the occurrence of non-territorial solitary wolves, which set up temporary lairs during the breeding period around Shurkul Lake. This appears to be due to the abundance of forage and protective conditions for them. According to herders, in March 2009, single wolves were observed attacking a herd of domestic animals. Out of a herd of 40 cattle belonging to Adilbek Zhumaniyazov, a shepherd, 6 cattle were eaten by a wolf. The victims were 2 cows with calves, two cows had their bellies gutted and another had their udders torn off. A single wolf attacked for 15 days at 2-day intervals.

We also conducted a questionnaire survey among shepherds of Kegeyli and Chimbay districts of Karakalpakstan (Aspantai, Shakaman). Based on the results of the questionnaire analysis, it was found that there were also wolf attacks on domestic animals here. According to Nurzhanov Khozhagaliy, head of the farm "Zhakhan" located in Kegeyli district, during 2010 it was noted that 15 foals were eaten by small packs of wolves (20-30 animals in total). According to oral reports from residents of the farm "1-Maya" in Chimbay district, 2 donkeys were eaten by a pack of wolves (9 animals) in August 2010. According to the data of resident Ensepbaev J., who lives in Shakaman settlement, 2 foals were eaten by wolves on 21 January 2011. Also on the territory of the farm Kuralpa in Karauziak district in February 2011 two wolves attacked cattle of local residents. According to the shepherd of Shakaman Rasbergenov Khozhabay, in 2004, in the summer of only 44 days about 200 cattle (mainly foals, sheep, etc.) were eaten by wolves.

In addition, according to questionnaire data, we found that there were frequent cases of wolf attacks on livestock in Uchsai village in Muynak district. For example, only in 1994-1995 wolves ate 2 cows, one of which was pregnant, and in 2002 wolves ate 5 cattle, including 3 cows and 2 bulls, in the household of resident Sarykulov Jenis.

Wolves everywhere in the territories we studied eat rodents. According to our observations and, according to literature data, rodents occupy the next place by occurrence in the annual diet, which is 24% [93, p.326]. According to Bibikov et al. (1985) their occurrence in the diet usually ranges from 2-3 to 10%.

It was noted that in the territory of the Southern Priaralie the following rodent species prevail in the wolf's diet: big, red-tailed and comb gerbils, yellow, small and thin-toed gophers, marmosets, occasionally lamellar-toothed rat, muskrat, house mouse, noon vole and blindfoot. Among birds, larks and field sparrow are found in its food. Of reptiles - turtle, lizards and snakes, various beetles and locusts are eaten. Of plants, green cereals, fruits of elk and mulberry trees are often eaten. The wolf occasionally preys fox (0.9%), corsak (0.8%), steppe polecat (1.0%) and long-eared hedgehog (0.9%). Of ungulates, adult wild boars (8.1-14.5%) and piglets (2.6-35.5%) are of great importance in the wolf's diet [93, p.326].

Thus, we note that the list of wolf food varies by seasons of the year throughout the study area of the Southern Priaralie. However, the predator's diet is similar in general character, and the difference for each season is determined only by the assortment of local food resources available at this time of year.

In winter, wild ungulates and rodents dominate the predator's diet, and it often attacks sheep and goats, birds and eats carrion. In spring, saiga, wild boar and hare are also of primary importance in the wolf's diet, but at this time it also preys on sheep. In summer, wolves congregate in areas with accessible water holes, where sheep flocks are kept and saigas breed. In summer, wolves congregate in areas with accessible water holes where sheep flocks and breeding saigas are kept.

In autumn, in addition to ungulates, rodents and hares are often found in the food of this predator, and the proportion of prey birds, reptiles and insects decreases markedly. Wolf attacks on wild ungulates become much rarer in the years of mass reproduction of hares. In recent years, according to our observations and population surveys in the central part of the region, hare numbers have remained high. We have also found numerous hare remains in wolf faeces.

During the period of breeding and rearing of young, the wolf's diet changes from large animals to small animals. This is due to several reasons:

1.   As a consequence of a sedentary lifestyle tied to a den;

2.   During the non-gregarious lifestyle, there is a dramatic decrease in the availability of large prey for solitary individuals;

3.   Wolves during the period of lactation and rearing their young have a need for diverse and nutritious forage (rich in various trace elements, minerals and vitamins), which are small animals and vegetation. The latter aspect is more characteristic not only of the wolf, but also of other terrestrial predators.

One of the important sources of wolf food is carrion. The presence of carrion in the lands is related to the natural death of various wild animals, to the creation of food reserves by the predator itself during the period of prey surplus and as a result of human activity.

It is known that in late winter and spring, before the emergence of young animals in wild animals and the beginning of grazing, when the remains of fallen animals begin to melt from under the snow, carrion is the most important food for wolves. In addition to accidental discovery of carcasses of fallen animals, in which wolves are helped by birds, wolves repeatedly visit the places of previous successful hunts and eat the remains left by them during the period of abundant food.

According to our observations and oral reports of shepherds of Kegeyli (Aspantai) and Chimbay (Shakaman) districts, wolves inhabiting their territories rarely feed on carrion due to the fact that there is an excess of forage base here. In the Northern part of the Southern Priaralie in the territory of Ustyurt under conditions of jute and ice, carrion is also included in the wolf's diet.

The daily diet of the wolf remains controversial according to different literature data. Specialists give the following examples: one pack (7-8 wolves) can eat the meat of a whole horse carcass, two wolves can eat at a time a boar carcass weighing 30-40 kg [18, p.123-193]. According to our data and oral reports of residents of the farm "1-May" of Chimbay district in August 2010, a pack of wolves (9 individuals) ate 2 donkeys (adults). The assumption that hungry wolves can eat enough food is doubtful, but starving wolves can eat at one meal an amount of food exceeding the usual daily norm [8, p.605].

Wolves are the largest and most skilled hunters. When they catch their prey, they usually cut open their stomach to get to the liver and other internal organs. According to the authors "If they are really hungry, wolves will, eat all the organs - bones, skin, fur, whatever is available - and as fast as they can." If they normally eat well, they will first, eat internal organs and meat, then the rest, and later come back for what's left. They will not, eat the horns, very large bones, or rows of teeth of an adult victim." [151, c. 472].

Many scientists believe that wolves eat large amounts of wool and fur, which is confirmed by analyses of wolf faeces. According to our data, the stomach of a wolf trapped in the Aspantay area in May 2011 was analysed and found to contain scraps of fur, small ribs, and pieces of meat from the victim. In total, the mass of the stomach with its contents weighed 1 kg. The stomach contents separately totalled 300 g. The scientists also noted that cannibalism or eating their dead kin is probably common among wolves. A wounded wolf always avoids its fellow wolves and joins them when the wounds are somewhat healed. During the rut, cannibalism in wolves is also possible with perfectly healthy individuals. Sometimes cannibalism of the mother to incompletely developing pups is observed [34, p.500]. According to our observation and survey data, cannibalism was not detected in the studied region.

Summarising the above mentioned and according to our observations, the highest percentage of forage resources of anthropogenic origin (domestic animals, cultivated plants, household waste) in the predator's diet is inherent to the population in the zone of the Central part of the Southern Priaralie, where settlements are located, the lowest percentage belongs to the Ustyurt population of the wolf.

Thus, it has been established that domestic animals dominate in the diet of wolves, which is apparently associated with the deterioration of the forage base of natural origin, including the drying up of most lakes and due to changes in the hydrological regime in the Southern Priaralie region. This is confirmed by the fact that a greater concentration of wolf packs is observed around the preserved small lakes, where the forage diet is more stable.

3.3. Reproduction and sex and age structure

To date, the reproductive processes in the wolf population have been studied quite sufficiently, but in the conditions of the Southern Priaralie, the study of this issue requires more recent information. As is known, the stages of reproduction in different geographical populations differ markedly in timing and intensity. Wolves are monogamous animals, forming pairs for many years. In case of death of one partner, it finds another pair.

According to numerous observations, it has been established that she-wolves of the same pack have heat at different times (early and late). Early heat is observed in adult females, while later heat is observed in young she-wolves. This is due to the fact that late heat in young she-wolves is an adaptation of the species and can compensate for the loss of the total wolf population. In addition, there is another adaptive ability, which is observed at high population densities and lack of free areas of territory not participation of young but sexually mature females in reproduction [147, p 834].

According to the literature, the rut in each pair of wolves lasts about a month. The rut consists of the pre-rut period, i.e. the pre-fluctuating state of the female and her own rut, to which mating is timed [8, p.605].

According to Palvanazov (1974) it was found that the rut in wolves in the Aral Sea region is observed in the first half of January and lasts approximately two weeks. According to our questionnaire data and oral reports of Khozhabay Rasbergenov, a shepherd (Kegeyli district), the rut in the Central part of the Southern Priaralie begins approximately in early February and the rutting period ends in the second decade of February. The same information is given by Ryabov K.A. (1980), where, according to his data, in the Central Black Earth Region, the onset of heat in she-wolves occurs from 8 February to 17 March. Wolves (females and males) from different packs in the number of about 15 individuals gather in the chasing pack. During seven years (2009-2016), we recorded a yearling pack of about 15 individuals in early February on the territory of the farm Aspantai (Kegeyli district).

In all cases, the chasing flock was observed in the morning and afternoon on rookeries and at the crossings. Deviations of these dates are also known, which are indicated in the works of specialists [93, p.326; 111, p.75]. Sometimes there is an earlier heat in she-wolves (late December - early January). But such deviations were not observed in our region. The considered cases of rutting occurred in the same period.

The timing of the rut varies depending on the habitat. According to Bibikov D.I. et al. (1985), the earliest onset of rutting is observed in December (southern Caucasus, Ukraine, Kazakhstan and Central Asia) and later in late December (southern regions of

European parts of Russia and Ukraine). Although according to our observations on 18 March 2011 in Shakaman (Chimbay district) shepherds caught a female caught in a trap at the age of more than two years (overyarok). Upon examination of the individual, it was found that the female was pregnant (pregnancy was approximately 30 days). The total weight of the female was 33 kg, while according to Palvanazov (1974) the average weight of a non-pregnant female desert wolf should be 20.3 kg. Pregnancy of the she-wolf lasts 62-65 days. For the region of Southern Priaralie it is characterised by 62-75 days. The birth of cubs in the region of Southern Priaralie falls on the beginning and middle of April. Cubs are born blind and become sighted on the 9th-12th day. For more than a month the cubs feed only on their mother's milk, then gradually get used to meat food, initially chewed and regurgitated by their mothers, and partly plant food. In early June, they begin to regularly visit watering places located 50-100 metres from the den. At three months of age, wolf cubs reach the size of an average dog and require more food. Until October, i.e. until the age of 5-6 months, young wolves stay in the den area and live off the prey brought to them by the parent pair.

According to experts, the number of pups in a brood varies from 2 to 9, more often 6-7, with an average value of 5.9. Fecundity of females depends, first of all, on food supply during the breeding period, as well as on the age of the she-wolf, density of its own population and intensity of human extermination [18, p.123-193; 8, p.605, 113, p.1-3; 13, p.70-71, 130, p.20, 140, p.7-9].

According to questionnaire data, in 2007, a den with 10 pups (5 females and 5 males) was found near Lake Tabankul in Chimbay district (former state farm, Pravda). The location of this den is shown in Fig. 20.

Fig. 20: Discovered wolf burrow in Chimbay district

In the Karauziak District, a den was also discovered in 2010 on the territory of the Nurumtubek Forestry in Nurumtubek, which was inhabited by two adults and two pups. It was established that both pups were males (Oral report of gamekeeper Atashov A.).

Based on long-term observations of the sex and age structure of this population, it was established that in the south of the Aral Sea region, the average fecundity of the wolf is up to 5.9 pups, as confirmed by the

by literature data. Brood size varies from year to year. Mortality remains high throughout the first year of life. On average, 1-2 out of 6-7 pups born in a brood reach sexual maturity (2 years of age), rarely 3. A large percentage is attributed to mortality of young pups. According to our studies, embryonic mortality is low, averaging 5%. According to Bibikov et al. (1985) in the northern part of the Aral Sea region embryonic mortality can reach up to 13% due to harsh climatic conditions and scarce forage base. It is also possible that fecundity depends on the age of she-wolves, on the population density and on the intensity of human extermination.

The sex and age composition of the model Aspantai-Shakaman wolf pack was studied over the period 2009-2016. Analysis of the ratio of males to females (**d:$**) in the wolf population showed that initially there was a slight predominance of males (5:3). A sharp decline due to intensive poaching by shepherds and the local population was accompanied by an increase in the proportion of females in the offspring (1:2.5). In this way, the populations sought to compensate for their losses, which were observed in 2011-2012. Further, the composition of the studied flock consisted of 10 individuals, including 1 maternal male, 1 maternal female, 3 perejar and 3 incomers, 1 adult male, 1 old male (Fig. 21). This type of pack is characteristic of all wolf distribution areas.

The sex and age composition of the wolf population is formed under the influence of fertility and mortality parameters. If the wolf population size is stable every year, the birth rate is equal to the mortality rate from year to year. From each parental pair of wolves by the end of their life there remains also a pair of breeding individuals. Observations showed that in the following years the pack composition changed in additions of individuals and transition of adults to newly forming packs, as well as mortality of individuals.

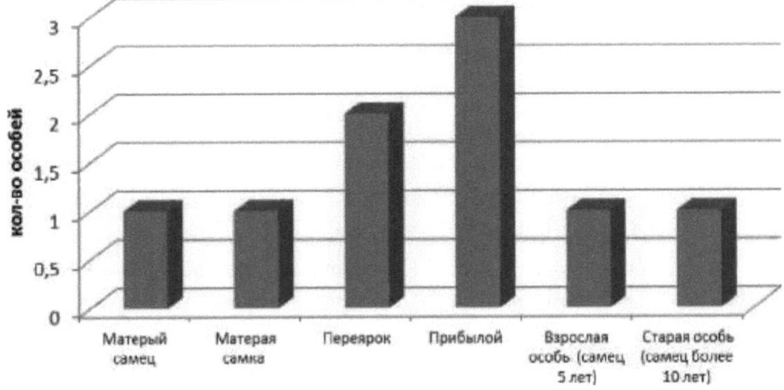

Figure 21. Average indices of sex and age structure of Aspantai - Shakaman wolf packs

45

According to specialists [111, p.75], a she-wolf takes part in three breeding seasons during her average life span (4.16 years) and brings about 15 pups. Up to sexual maturity 87% of the litter does not reach or must die. Litter mortality may not be evenly distributed over the years. There may be alternating periods of increasing numbers with periods of decreasing numbers. The longevity of an individual can also periodically increase (up to 5.3) and decrease (up to 3.6 years), changing the yield of young per she-wolf and thus changing the course of population dynamics.

This is confirmed by the data of Ryabov (1980). As he noted, a wolf pack is a wolf family including mothers (parent pair), newcomers (young up to a year old) and 2-3 perejarks (young of the previous generation). Occasionally, a pack may include single individuals from earlier litters, usually males. In 2003, one male Pereyar from this flock fell into a trap and lost the second toe of the front left foot. Further we studied traces of male activity (trolling, rookeries, feeding, etc.) according to this trait. Over the next 2-3 years, observations were made on the flock, which was increased to 15-20 individuals. From these flocks, sexually mature males and females of Pereyarka formed new flocks with individuals from other flocks.

Thus, the number of each flock was controlled by the individuals themselves. In 2007, at the end of June, the shepherd Rasbergenov Khozhabay found a den (at the coordinate N-43° 15ı, E- 59° 19ı 21) with 10 pups. The sex ratio was 1:1.

Repeated observations of wolf tracks of the Aspantai-Shakamanskaya pack revealed that the pack consisted of a parent pair (1 maternal female and 1 maternal male), 1 mate, 1 adult male (4-5 years old), 1 female overjack (more than 2 years old). From these packs, a wolf aged 4-5 years was trapped by local shepherds in May 2011. This confirmed the conclusions that there may be males from earlier litters in the pack.

In 2011, on 18 March, a pregnant female Pereyar from this flock (over 2 years old) was also caught in a trap. As a consequence, apparently, there were no more breeding females in this flock, so there was no litter, and the mother female did not participate in breeding due to old age. With regard to the latter two indicators, even in large samples pooled over a number of years, the ratio of one age group to the previous one fluctuates for random reasons. This is especially noticeable in small samples, where there are so few adults that it is necessary to smooth out random fluctuations by equalising the age series using the least squares method. A decreasing geometric progression is found that best fits the actual number of animals in the six age groups. The measure of best fit is χ-squared, tending to minimise at the desired value of mean survival for the age interval from 2+ to 7+. The mean maternal age varies similarly to the survival rate. However, the relationship between the two is not straightforward. In addition, animals older than 8 years of age are included here. Survival rate can be used to estimate the intensity of extermination, and mean age can be used to estimate the potential replenishment of the population with young.

It should be noted that a change in age structure may be the result not only of the intensity of removal, but also of any change in total abundance. For example, during the growth phase, each successive generation is more abundant, so the older the age group, the more sparse it is, but not as a result of intensive killing, but because of its sparseness at birth [33, p.362]. The average age and survival rate will be underestimated. Obviously, this indicator will live up to its name at stable (high or low) numbers. In any case, it turns out to be useful in interpreting the age structure of the wolf and its analysis.

According to A. Bondarev (1979), the sex ratio of wolves is an indicator of population fluctuation. To determine the sex and age structure of wolves, studies were conducted in Muynak, Takhtakupyr, Kanlykul and Ellikkala districts of the Southern Priaralie (Republic of Karakalpakstan) during 2009-2016. The sex composition of wolves harvested by amateur hunters was analysed (13 individuals, including 6 adults and the rest pups, according to the questionnaire-questionnaire method). The age of adult animals was determined to the nearest year by the number of annual rings formed in the cement layer of teeth during winter. Arrived wolves were determined by the ratio of canal width to canine width [112, p.44-45].

Under the constant pressure of hunting and other anthropogenic factors, fluctuations in wolf numbers are determined mainly by the number of females in the population. According to questionnaire data and interviews with residents of the surveyed territories, in case of shortage of males, wolves mated with dogs. This was observed in the territory of Kanlykulsky (4 wolf individuals were observed, among which there was a wolf of mottled colour) and Takhtakupyrsky districts (two wolves out of 6 individuals were of black colour). This is probably due to the fact that the number of sexually mature males in the pack in these areas decreased. According to experts, the main reason for the appearance of wolf-dog hybrids in nature was a significant decrease in the number of wolves as a result of human persecution, which was accompanied by the disintegration of wolf packs and disruption of the sexual structure of predator populations.

In the semi-desert zone, pack fragmentation is caused by the nature of summer feeding of predators. From April to August, wolves in the sandy desert of northwestern Kyzylkum live at the expense of rodents. With this type of feeding there is no need to unite into packs, as it usually happens when hunting ungulates.

In Karauziak district, a den was found in the territory of Nurumtubek forestry, inhabited by two adults and two pups. It was established that both pups were males. According to experts, based on the analysis of the collected long-term material from the entire CIS region, it was revealed that males prevail among wolf cubs or they make up an equal share with females [8, p.605].

According to Palvanazov M. (1974, 1990), the average brood size in the Southern Priaralie is 6 pups, and deviations are rare. Mortality remains high throughout the first year of life. On average, 1-2 out of 6-7 pups born in a brood reach sexual maturity (2 years of age), rarely 3 owls. Mortality of young animals accounts for a large percentage. According to Heptner W. et al. (1967) infant mortality in wolves is 60-80%. The same information is given by Bibikov D. et al. (1985). The brood size varies from year to year. This is related to the availability of basic fodder and the impact of

anthropogenic factors. According to specialists [88, p.21], in spring in April-May months, an average of 5 wolves were found in seven wolf brood dens. According to long-term data, it is established that if forage availability
(wild ungulate, hare-tolai and rodent numbers) was low, the average number of litters decreases to three pups.

The age structure is characterised by the death of mature individuals, which does not always lead to the death of newcomers, as parental functions can to a large extent be performed by pereyarkas and young wolves that did not participate in breeding this year. However, the activity of wolves is controlled by them much less strictly. The young wolf-raisers are least successful in teaching the cubs the subtle features of interaction with the environment, including, formation of social and sexual preferences.

According to our observations and interviews with the local population, there are two wolf family packs in Uchsai and Kyzylzhar tracts in Muynak district with a total number of about 15 animals (4.1 wolves per 1,000 km^2). The most numerous wolf pack was observed in Uchsai tract, with 7-8 animals in October 2009.

When the number of wolves increases under conditions of food supply, the size of packs grows up to a certain limit, after reaching which there is only an increase in the number of non-territorial animals. At the same time, mortality increases, birth rate decreases, and only females participate in reproduction. Due to these processes, the population size stabilises at a certain ratio of the percentage of territorial and non-territorial animals. When forage resources are reduced, wolf numbers decline, which is reflected in a decrease in pack size.

Population depression occurs due to mortality of individuals from exhaustion and from increasing intraspecific competition. Pups and low-ranked animals die primarily from malnutrition. The remaining wolves are forced to expand their hunting grounds in search of prey and thus come into contact with neighbouring packs more frequently. As a result, clashes between them become more frequent, often resulting in fatalities.

In the course of expeditionary studies and according to survey data, the frequent occurrence of solitary wolves was established. According to literature data and after consultation with specialists from Zhitkov VNIIOZ (Kirov, Russia), we came to the conclusion that in the conditions of the Southern Priaralie with insufficient snow cover wolves prefer not to unite in packs. There is another hypothesis, which points to the fact that at high wolf numbers intraspecific competition occurs. Due to competitive relations between individuals, lower-ranked wolves are driven out of the pack, which in turn become non-territorial solitary individuals. In years of low wolf numbers there are associations of 2-3 wolf packs, which number reaches up to 20 individuals. Such a fact was observed on the territory of Akbetkei archipelago.

Thus, in populations with undisturbed self-regulation there is a saturation limit, beyond which growth is impossible. Discussion of the results of the wolf study in comparison with other parts of the studied territories in this aspect is quite reasonable and expedient, since the principle mechanisms of population regulation for the species are the same [33, p.362]. With increasing numbers of wolves in their populations there is a decrease in fecundity and an increase in embryonic mortality. At decreasing numbers - on the contrary, fecundity of females and survival rate of young increase. At high population density, intrapopulation regulation is manifested through changes in fecundity.

The ratio of different sex and age groups in populations determines the ability to reproduce at a given moment and reveals the trend of population changes. An increase in the proportion of young in populations is a result of favourable breeding conditions, an indicator of population recovery from depression and growth of animal numbers.

In years of low numbers of animals, spatial changes in the predator population are observed, manifested in significant separation of individual groups - packs of wolves from each other and their confinement to populated areas, especially around small lakes.

The variety of landscape and climatic conditions, diversity of biotopes, different anthropogenic load on the wolf population and their habitat determine the formation of certain features of territorial distribution, movements, daily and seasonal activity, feeding patterns and some other features of wolf ecology in the Southern Priaralie.

3.4. Burrows and shelters

Studying the ecology of mammalian predators in different regions is of great scientific and practical importance. Changes in the natural environment and direct persecution lead to a sharp decrease in the number of wild animals.

For wild animals to live and breed properly, they need natural shelters or places to raise their families and offspring. Not everyone knows how important protective conditions are for wild animals. In order to preserve wild animals, it is necessary not only to protect them from extermination, but also to treat with care the natural conditions in which they live. Lack of shelters, lack of conditions for brood nests not only reduce the number of animals, but also lead to their complete extinction in the area. People in the process of economic activity often unknowingly violate and sometimes completely destroy animal shelters. The most common animal shelters are burrows. According to researchers' data, from polar seas to hot deserts and tropics inclusive, they serve as temporary or permanent dwellings for most mammals [93, p.320; 137, p. 72- 84; 61, p. 62- 68]. However, these animals use them in different ways. Some spend their whole life in them, others inhabit them only for a certain season, for others it is a place of cubs' upbringing, and there are such animals that do not live in burrows themselves, but hide their offspring in them. The main features of these shelters are that they do not exist in nature in a ready-made form - animals create them themselves. However, not all burrow dwellers are able to dig them. Many settle in abandoned burrows, prepared by true diggers, or even displace their rightful owners from there.

Burrows are long-lasting shelters that can be used for decades by many generations of animals. Their structure is extremely diverse. The simplest burrow is a straight tunnel sloping downwards and ending in a nesting chamber, sometimes lined with some soft dry material. More complex burrows are endless multilevel underground labyrinths with a multitude of roundabouts and dead ends, entrances and living chambers. Burrows often change owners, and each new owner can modify the shelter to his taste. Burrows have their own microclimate. They are warmer in winter and cooler in summer. Apart from mammals, arthropods, snakes, toads and birds also settle in burrows. A new settler together with the shelter receives ectoparasites of the previous host, ticks and fleas, sometimes leading to the spread of dangerous infectious diseases [107, p.84-109].

In the course of our study, the features of natural shelters and dens of wolves in the Aral Sea region were studied. The results of the study revealed that the dens and shelters of these predators were adapted in their own way to the anthropogenic pressure and distributed in different biotopes in the changing environmental conditions of the Aral Sea region. The structures of burrows were also studied.

Wolves choose den sites more often in the most remote places rarely visited by humans. In Central Asia, the animals make their lairs in thickets of tugai, reeds or in mountain gorges. In the north - in forest islands, bushes and sometimes simply on the banks of rivers, streams and rarely lakes [75, p.8-50].

The design, size and location of wolf dens vary and depend on specific habitat conditions. The general requirements of wolves for den construction are: hidden, low human-visited location, availability of a water source and reliable protection from unfavourable weather, and availability of food. The approach to the den is more often hidden - there are usually no rotting remains of meat, claws and other objects that give away its proximity [8, p.605].

Wolves rarely dig dens. They prefer to use other people's shelter, and if they cannot find such a shelter in a suitable place, they often bear offspring in the den. However, sometimes they have to dig their own burrow. Their burrows are simple and more often with one exit [107, p.84-109]. In addition to primary and secondary dens, there are permanent denning sites on the family's territory, which wolves use both during and after the breeding period. The denning sites combine good protective conditions [8, p.605].

During expedition trips in 2015, we discovered wolf dens in the Aspantai-Shakaman model area of the Kegeli District. There were several lairs in this area, which were located 23 kilometres from Aspantai village. Sheep flocks were located nearby. The first den was destroyed by shepherds back in 2005, from where 11 wolf cubs were taken.

It is known that the conservatism of wolves in the choice of den is great - even after repeated destruction of wolf cubs and mature animals in the same dens, cases of repopulation of ruined shelters were observed [140, p.7-95]. This is confirmed by our studies. Thus, we found a burrow, which was dug at the bottom of the old Amu Darya riverbed (at the coordinate N 43^0 01^I .50, E 59^0 19^I .21). The soil here was loose. The burrow was arranged simply. It had one entrance. The burrows were absent. The width of the burrow entrance was 60-63 cm, length 500-600 cm, height 34 cm.

Another burrow was found 50 metres from this den, with an entrance hole 50-60 cm wide, 520 cm long, 33 cm high, and the nest, located in the middle of the burrow, had the following dimensions 490, 80 and 105 cm, respectively. The nest was half filled with dry grass bedding. The burrow had an emergency exit.

A few metres from the burrow there was a collector-drainage network. Thus, we note that in the conditions of the Aral Sea region wolves used the same territories for breeding their offspring for many years.

Wolves dig their dens, sometimes occupying badger and fox dens. During the survey we found several badger dens occupied by wolves, as well as lairs and rookeries in dense reed thickets around the lakes Mezhdurechye (N 43^0 55.48.71., E 59^0 25.39.56) and Zhaltyrbas (N 43^0 .52.087., E 59^0 .42.893.), as well as under the dzhida, in thorny thickets of jingil on the territory of the Kazakhdarya forestry hunting farm.

During expedition trips to Sudochie Lake, wolf dens were also discovered in the gorge (at coordinates N 43^0 .52.32.43, E 58^0 .36.40.10.) at an altitude of 134 metres above sea level. This wolf burrow in a ravine on the shore of Sudoche Lake (at coordinates N 43^0 .58.40.49, E 58^0 .53.40.49) was located 1.5 metres above the ground.

The second burrow was found about 15-20 km from the first. Excavation of the burrow to study its structure showed its relatively simple structure and large size. It started with a common entrance and was divided into two independent snouts. One was 800 cm long, 60 cm wide at the bottom and 40 cm high at a depth of 3.6 m and ended in a dead end. The second was 100 cm long, its nesting chamber was at a depth of 3.2 m and had a diameter of about 50 cm with a dome-shaped vault almost 80 cm high. No bedding was found in the nest. The burrow was dug in sand and had an almost horizontal profile.

As noted by Kozlov (1955) the places chosen by wolves for denning fulfil the following requirements :

1. Relative remoteness of the site (areas rarely visited by humans, although they may be in close proximity to human settlements)

2. The relative secrecy of the approach to the den

3. Proximity to a body of water.

Based on the data we collected, we identified 2 main types of lairs: formed and weakly formed burrows in sheltered areas, which have transitional forms (Table 1).

Table 1

Distribution of wolf lairs depending on relief exposure

Types of construction lairs	Number of norms surveyed	Location		
		Plateau Ustyurt	Riparian and reed beds	desert-steppe
Formed burrows	8	2	3	3
Weakly formed burrows in sheltered areas	2		2	

This agrees with the data of Palvanazov M. (1974, 1992), where he noted that in case of weak anthropogenic factor wolves make dens most often on the slopes of ravines and in hollows overgrown with saxaul and comb. Thus, according to the oral report of G. Turemuratova, Candidate of Biological Sciences, during the expedition trip in 2010 to the territory of the dried-up Aral Sea bed, a dug-out den was found on the slope of the slope of the dried-up territory of the sea. In the floodplains of the Amu Darya River, where people often visit, wolf dens are located in hollows, riparian and reed thickets. Thus, in the territory of the Southern Priaralie it has been established that wolves prefer to make dens around small lakes and water bodies, as well as in dense reed thickets, and in desert zones they prefer to settle near artesian wells and settlements of shepherds. This is one of the peculiarities of the ecology of the wolf inhabiting the territory of the Southern Priaralie.

3.5 Social organisation and behaviour

Behaviour is one of the most important ways of active adaptation of animals to a variety of environmental conditions. It ensures the survival and successful reproduction of both the individual and the species as a whole. The study of animal behaviour has consistently attracted widespread attention for many reasons. Information on animal behaviour is essential for understanding animal ecology (lifestyle), which, in turn, contributes to the development of conservation and environmental management.

Many works of famous scientists [133, p.487; 138, p.856; 21, p.40; 81, p.100; 135, p.568; 76, p.520] are devoted to animal ethology. Each of them represents a rather original presentation of the basics of the science of animal behaviour. The behaviour of predatory mammals, including the wolf was also covered in the works of such famous scientists as Manteifel P.A. (1947) Krushinsky L.V. (1986) and Korytin C. A. (1976).

Different approaches can be used to study animal behaviour. Behaviour can be considered from the point of view of its formation in evolution, from the point of view of the benefit it brings to the animal, you can also consider its psychological or physiological mechanisms. The choice of approach is determined by what exactly you want to know about animal behaviour [76, p.562].

At present, a great deal of material has been accumulated that characterises behaviour as a set of different forms of adaptive activity of predators. Animal behaviour is infinitely diverse in its forms, manifestations and mechanisms. The current systems of classification of behaviour are manifold, as the number of criteria that can be used as its basis is practically limitless. The classification of Dewsbury D. (1981) divides behaviour into three main groups - individual, reproductive and social [39, p.105]. Manifestation of all forms of behaviour is influenced by daily, seasonal and other biological rhythms [25, p.320]. Studying the behaviour of wild animals under the influence of anthropogenic factors on them becomes necessary in the face of the inevitable, ever-increasing onslaught of civilization on nature.

The wolf is characterised by its great ecological plasticity and with a highly developed psyche. This allows it to successfully resist all the variety of ways of fighting it. For a long time, the behaviour of animals, in particular wolves, was defined in terms of their instincts and conditioned reflexes.

Many scientists have come to the conclusion that animals have the ability to analyse the situation, draw certain conclusions and predict events. Even Zworykin N. A. (1937) noted high mental activity of wolves. It is especially clear during collective hunts of predators on different animals. Wolves disperse to different areas to search for prey, signal the results of the search, surround the detected prey, ambush it and catch up with its lurking brethren or in areas (on ice, in deep snow, etc.) where it is easier to take it.

Wolf behaviour is extremely complex. This animal sometimes leads such a secretive way of life that its den or burrow, located near populated areas, cannot be detected for a long time. The wolf is known to be very observant and can find carrion by the cry of crows and magpies, for example. It precisely determines its behaviour in accordance with the situation [91, p.351]. The hunting activity of the wolf and the choice of its victims depend on the season and the family status of the wolf (whether it is in a pack or not, young or old).

An important characteristic of the species is the presence of such a form of collective survival and interaction as a pack. A pack is essentially a family consisting of a parent pair and young wolves up to two or three years old [132, p.72].

In the Aral Sea region, the study of wolf ethology was devoted to the works of Palvanizov M. (1974, 1990). The results obtained in the course of our study show that the studied wolf population consists of pack-united individuals that use certain territories of indigenous areas and single animals that are not part of a pack and are characterised by high mobility. The size of the habitat area of the wolf population in the Aral Sea region is determined by a certain landscape and varies in different areas. The size of the territory and the density of one pack depends on the food supply, the availability of shelters and the location of water sources.

A strict hierarchy is maintained in the pack. Its basis consists of an alpha male, an alpha female, several low-ranking wolves of both sexes, among which the beta male may stand out, and pups outside the hierarchy. This is supported by our research. On the territory of the model plot (f/fx Aspantai - Shakaman), wolf packs live in family groups, the basis of which are - a pair of mature wolves with the litter of the current year (arrivals), offspring of the previous year of birth (overjunks), as well as one adult male (5-6 years old), one old male (12-13 years old).

According to our observations, wolves belonging to the same pack-family often hunt in a common area alone or in groups of 2-4 individuals.

In addition to wolves living in packs in certain territories, there are wolves roaming, "stray" and non-territorial. These are, as a rule, individuals who are over the age of overyark (there are both overyarks and incomers, as well as old animals), driven away by mother wolves and have not found a free area.

In the studied wolf pack, seasonal changes in family and territorial relations have been revealed, one of which is the brood period. The brood period begins with the birth of wolves (April) and lasts all summer. The brood period is followed by the pack period. The pack period begins in autumn (mid-September) and lasts until the first half of winter (February). During this period, overjunks join the mothers and incomers. The pack, either as a whole or disintegrating in different variants, roams throughout the whole family-pack area and goes beyond its limits usually only to the territory free from wolves. During the rutting period, the pack disintegrates. During this period (late January, early February), newcomers are separated from the pack and live separately in the same pack area. The rutting pack is formed of a mother she-wolf in heat and males pursuing her - a mother she-wolf and usually a mate, sometimes a challenger from other wolves. The composition of a chasing pack can be more complex.

In total, the distance travelled by this wolf population within the pack territory is about 200 km. Based on the results of the study, it was found that daily movements averaged 82-87 km. At the same time, it was noted that sometimes individuals of the population can violate the boundary of their territory and cross the Amudarya River by "floating bridge" in the direction of Kungrad district. There were also cases when whole flocks (about 11 individuals) of the Kungrad population crossed the territory of the Aspantai-Shakaman population in this way in search of prey.

According to American zoologists R. Peterson (Peterson, 1977) and D. Mech (Mech, 1970). Peterson (1977) and Mech (1970), wolves cover their territory with a network of odour tags, but they mark border areas most intensively. Wolves intensively "mark" the boundaries of their territory with urine, faeces and vigorous "scraping". Thus, in May 2015, following the trail of the wolf pack of the Aspantai-Shakaman population, we counted several scrapes at road crossings every 5-6 kilometres, and in three places the wolf left excrement at a distance of 2-3 metres. The wolf travelled using motorways. On six occasions it urinated near jingleberry bushes around the road. Sometimes scrapes were found around the urinary point. The scrapes were either close to or a short distance (2 metres) from the urinary point (the largest scrapes measured 190 x 50 centimetres).

Thus, gregariousness and territorial movements of the wolf population play an important role in the stability of maintenance of vital activity and population size, as a behavioural response to changing environmental conditions of this species. According to our studies in the Southern Priaralie, the size of territories of one pack varies from 150 to 1200 km^2 depending on relief and landscape.

Seasonal fluctuations in the size of wolf habitat areas are also closely related to food availability, so in open landscapes where ungulates migrate widely and in winter food becomes scarce. In this case, there is a sharp entrainment of the area of the utilised territory. Examples are the Ustyurt Plateau and the Kyzylkum desert. A wolf pack occupying the indigenous area lives in family groups, the basis of which is a pair of mature wolves with the litter of the current year (new arrivals), offspring of the previous year of birth (pereyarka), and sometimes even older ones, apparently descended from the same mature in most cases males. Belonging to the same pack-family, wolves often hunt in a common area singly or in groups of 2-4; it is this gregariousness that is reflected in the analysis of wolf sightings and tracks. The number of individuals in a pack-family is determined by sightings of the largest number of individuals together in a given family plot and is verified by all information from the plot - the whole family may never be seen. The number of solitary non-territorial individuals is about 10% of the population, provided that the area is predominantly populated by wolves. They may join driven packs and sometimes become established, having driven out the "master" of the area, or replace one of the mature ones in case of his death, but more often they move to uncomfortable areas and in most cases perish. They are the ones who appear on territories not inhabited by wolves. At high wolf population densities, "stray" animals can account for up to 40% of the population, travelling far from their birthplaces, following herds of ungulates along the entire route of their migrations; sedentary wolves usually only accompany these herds within their areas [110, p.8-57].

So-called "buffer zones" are formed between wolf pack territories. Studies by Mech (Mech, 1977, 1979) in northern Minnesota have shown that buffer zones, i.e. border strips between pack territories, function as a reserve of wild ungulates. Wild ungulates inhabiting these strips are subject to wolf attacks to a much lesser extent than those living in the central parts of the patches. As a consequence, the inhabitants of the periphery of wolf territories survive longer. This phenomenon is due to the specificity of interspecific interactions between wolves. According to Mech (Mech, 1977), the width of buffer zones ranges from 1.6 to 3.2 square kilometres, with a width of 25% of the area of wolf pack territories. In our cases, the buffer zone varies within the same limits. This strip is visited by both neighbouring packs and it is here that encounters between them and in the course of a fight one of the wolves may be killed. Therefore, wolves usually do not linger in the buffer zone. The tension of their behaviour at the border of the territory is confirmed by a sharp increase in the frequency of tagging (2 times as compared to the centre). When ungulate numbers are high, wolves rarely hunt them in the buffer zone. Due to this, when ungulate numbers are depressed, their population density decreases primarily in the centres of wolf pack territories.

The shape of the root section is usually somewhat elongated, oval. Several zones or areas are distinguished within it (Fig. 22).

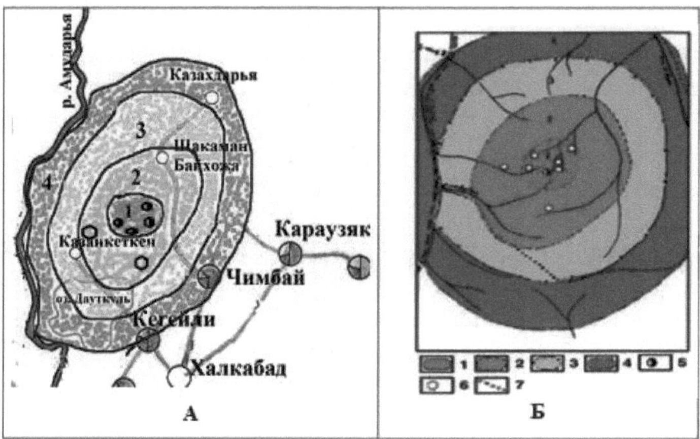

Fig. 22. Schematic of a wolf family pack territory

A - Schematic diagram of the Aspantai-Shakaman model wolf family and pack area;

B -Classical diagram of a wolf family pack territory (according to Suvorov, 2009)

1 - nesting area; 2 - summer brood (roost) area; 3 - hunting -

foraging area of the family flock; 4 - border zone; 5 - den; 6 - reserve burrows.

The nest site has an area of 4-5 km^2. It contains the den where the she-wolf cubs litter. There is also a brood area, which includes the nesting area and temporary lairs where the wolf cubs live after leaving the maternity den. In this area, the new arrivals spend the vast majority of their time during the brood period. This is also where the dens of mature wolves are located during this period. Pereyarka also visit this area. The brood area can be called the "home" of a wolf pack-family, which has an area of about 20 square kilometres. It is like the centre of wolf activity in the whole pack-family area.

The size of territorial indigenous sites of the Ustyurt wolf population is somewhat different from the Aspantai-Shakaman population. If the terrain does not change drastically, the houses exist for a very long time and serve many generations of wolves. It is known that even if the pack is completely wiped out, new wolves often choose the same territory for a home. In addition, there is the main hunting area. It includes the rest of the habitat area. The main hunting area, within which most of the wolves hunt during the brood period, has a radius of 8-10 km (with the centre approximately in the middle of the home range). There is also a border zone. It surrounds the main hunting area with a strip of 1-4 km, if this pack area neighbours the areas of other packs. Mature wolves rarely go beyond the main territory, marking and maintaining its boundaries. Pereyarka and newcomers are often in the border zone, and may even enter a foreign territory. Thus, the permanence of the boundary of the pack habitat is related to the presence of neighbouring packs.

In a single pack wolf population, seasonal changes in family and territorial relationships occur. One of them is the brood period. The brood period begins with the birth of wolves and lasts all summer. Arrivals live within the home range, first in the maternity den, then in the temporary den. Sometimes, especially towards the end of the period, the young make trips outside the house. The mother wolf hunts throughout the main area of the family plot. The she-wolf stays with the cubs for some time, then gradually hunts further and further away. Pereyarka at this time usually live on the edges of the pack territory, going to the border zone; sometimes, more often towards the end of the period, they "visit" the house.

During the pack period, which begins in autumn and lasts until the first half of winter, the pack, either as a whole or, breaking up in different variants, roams throughout the family and pack area. They usually leave the area only when it is free from wolves.

The above can be traced on the example of the Aspantai-Shakaman wolf population. During the study period, all daily and seasonal movements of wolves within the territory of the pack area of this population were traced (Fig. 23).

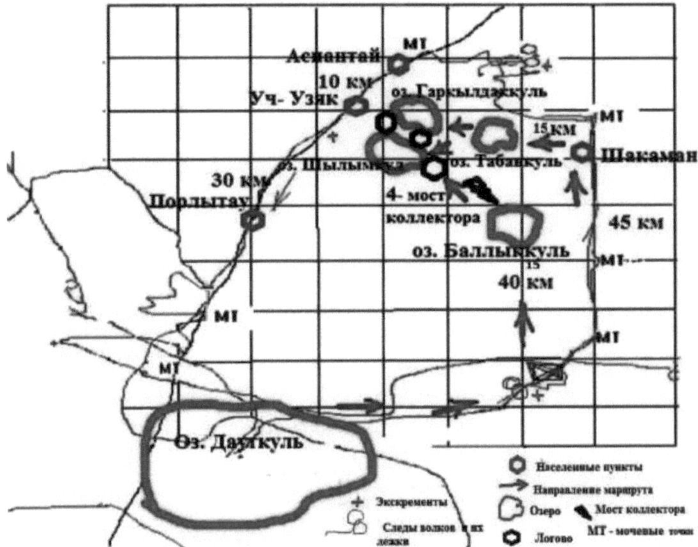

Fig. 23. Diagram of the wolf family pack route in the model plot
Aspantay - Shakaman

As can be seen from the diagram, the main core area is located on the territory of F/X Aspantai, Shakaman, Bozatau. In 23 km to the north-east of Aspantai village in the middle of the core area there is a nesting area. It contains a den, where the she-wolf is cubs. There is a brood area around Lake Shylymkul, which includes the nesting area and temporary lairs where the wolf cubs live after leaving the maternity den.

Each time after traversing their territory, they return to the area anew. The movement of wolves around the area is associated with the search for prey. The route of this population starts from the territory of Aspantai towards Uch-uzyak settlement (10 km), then it goes to Porlytau settlement (30 km), then it reaches Dautkul lake (30 km), then passing through the territory of the former Bozatau district (30 km) it can go either north-west towards Shakaman (45 km) or towards Balykkul (40 km).

Passing through the Balykkul area and the fourth bridge of the collector, they go to the breeding area and then return to the brood area in Shylymkul, Garkyldak kul. When travelling from Bozatau to Shakaman, they may return via Lake Tabankul (15 km) to the brood area at Shylymkul, Garkyldak kul. In total, the travelling distance of this wolf population is about 200 km.

The results of the study found that daily movements averaged 82-87 kilometres.

The wolves of the Ustyurt population can be distinguished from long-distance migrations caused by ungulate migrations (saigas, gazelles) when the pack moves in winter within its hunting area in search of food. First of all, the territory of Ustyurt is sharply characterised by its scanty forage base. The main prey of wolves in this territory is wild ungulates, including saigas, gazelles and wild boars. At the same time, they hunt small mammals (gophers, gerbils, gerbils, muskrats, etc.).

The territory of the Karakalpak part of Ustyurt, which is 7,000 square kilometres, is inhabited by approximately 6 wolf packs (each pack has an average of 7-8 individuals). The size of the indigenous area of each pack is about 1200 square kilometres. The large area of indigenous areas is associated with the scarcity of forage in the territory of Ustyurt. The wolf packs in this area build their breeding grounds near settlements or around water bodies and small lakes.

Both single and pack wolves were found near Jaslyk settlement, Karakalpakiya station, near Sarykamysh, Sudochie water reservoirs, Hydrometeorological station on Cape Aktumsyk.

According to Sludsky (1981), wolves accompany their prey, namely herds of saigas and gazelles travelling to the southern part of Ustyurt in September-October for wintering. As noted by Filimonov A. N. (1975, 1982), wolves "accompany" saiga herds, but often only as far as "their" borders, and then they are met by wolves in whose territory the migrants appear. After the saigas leave, wolves continue to live in their territories, feeding on the numerous saiga carcasses left by their successful hunts of these animals, as well as attacking livestock. We assume that some wolves (more often young, stragglers or lone wolves) accompany saigas for quite long distances, keeping behind the migrating herds. As saigas pass over a large area, they gradually accumulate a "plume" of such wolves, which increase in numbers when saigas arrive in a particular area. The same non-territorial wolves following saiga herds may inhabit wolf-free territories in the southern part of the Aral Sea region.

Thus, in areas with stable wolf populations, the territory of each pack is surrounded by several (5-6) neighbouring territories, i.e. the use of space is strictly regulated. Under such conditions, the size of wolf territories remains stable for many years. Territoriality is a mechanism of self-regulation of populations. Stability of spatial organisation is typical for the wolf population as a whole: territorial conservatism of the wolf is described in the literature [18, p.123-193; 20, p.29-38; 75, p.80-50; 40, p.90102]. This aspect should be taken into account in the development of measures to regulate wolf numbers, as resilience is a mechanism that ensures the preservation of the integrity of territories, not reduced, however, to simple avoidance of other people's odour.

3.6 Analysing wolf population dynamics

Population dynamics is an important parameter characterising the total number of individuals in a certain area during a certain period. When studying the issues of wolf population dynamics, most specialists paid special attention to the impact of various biotic and abiotic factors on their vital activity. It has been established that the effectiveness of various ecological factors as a mechanism regulating population size in ecosystems depends on the influence of anthropogenic impact [93, p.326; 94, p.76]. The conducted analysis of the relationship between the dynamics of the wolf population dynamics and the forage diet (gazelle) showed that this forage factor linearly depends (correlation coefficient R=0.4) on the predator density. But we note that linear trends of both species of animals tend to decrease (Fig. 24).

In natural conditions, where there are weak anthropogenic pressures, these processes manifest themselves year by year, depending on forage resources and protective conditions. In these cases, intra-population mechanisms of changes in abundance function, which are regulated by predators themselves and their competitors.

The forage factor, including one of the ungulates (wild boar) also plays a certain role in wolf nutrition. The analysis of the relationship between the dynamics of the wolf population dynamics and the forage diet (wild boar) showed that this forage factor linearly depends (correlation coefficient R=0.18) with disturbance moments on the predator density.

But we note that the linear trends of both animal species have a stable trend (Fig. 25).

Of small mammals, the tolai hare (*Hepus tolai* Pall, 1778) is an abundant species widely distributed in the Aral Sea region and plays a certain role in the wolf's diet.

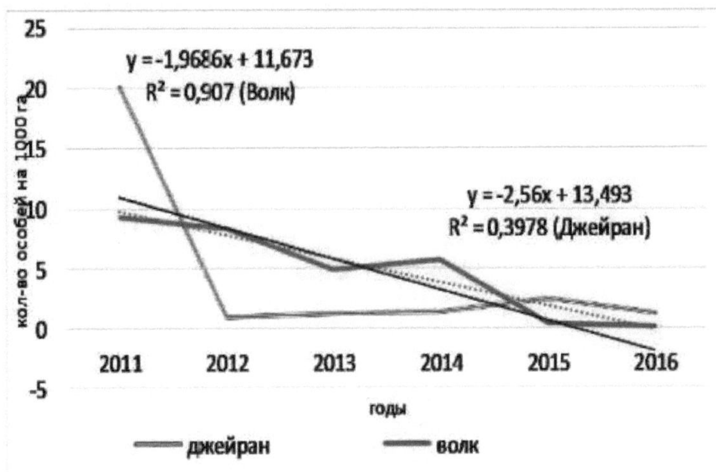

Figure 24. Dynamics of wolf population size depending on the forage object (Jeyran) for 2011-2016 (R=0.39)

Correlation analysis of the relationship between the dynamics of wolf population dynamics and the prey object (hare) showed that this prey factor depends very weakly but linearly (correlation coefficient R=0.06) with perturbing moments on the predator density.

But we note that the linear trends of both animal species have a stable trend (Fig.26).

The *saiga* antelope (*Saiga tatarica*) is the main natural prey species of the wolf in the Southern Priaralie. According to experts, wolf numbers are directly dependent on saiga numbers. Currently, due to the decline in saiga populations in Ustyurt, wolves are declining in numbers and migrating to the Central Priaralie [8, p.8-9].

An analysis of the correlation between the dynamics of wolf population numbers and the prey diet (saiga antelope) showed that there was a close correlation with this prey factor, which depends linearly (correlation coefficient R=0.71) on predator density.predator density. But we note that the linear trends of both species tend to decrease (Fig. 27).

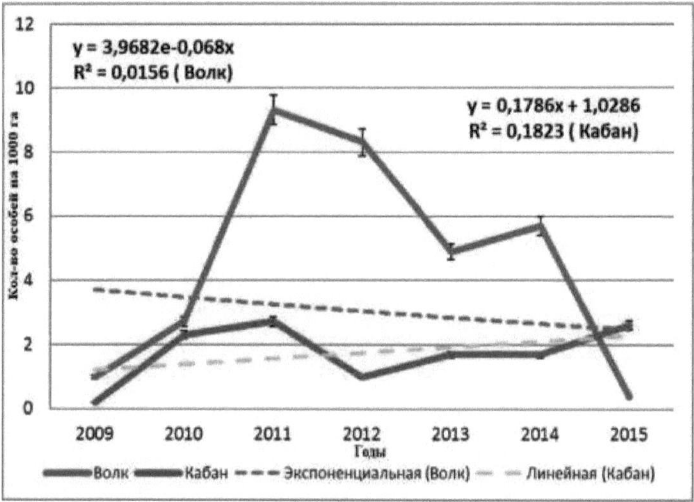

Fig. 25. Dynamics of wolf population size depending on the prey item (wild boar) for 2011-2016 (R=0.18)

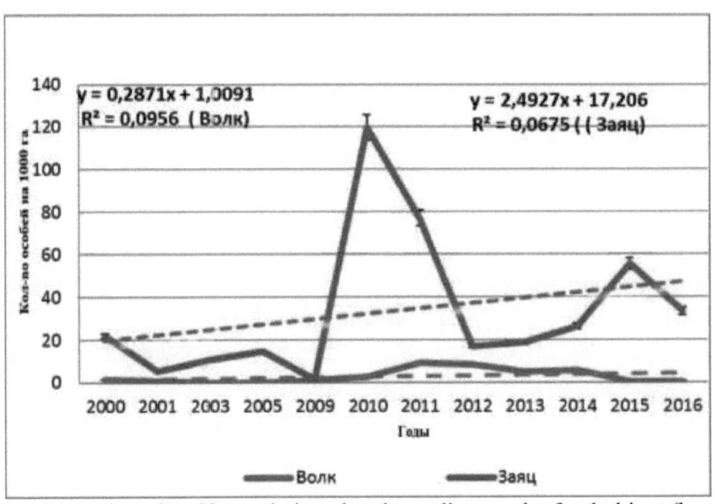

Fig. 26. Dynamics of wolf population size depending on the food object (hare) for 2011-2016 (R=0.06)

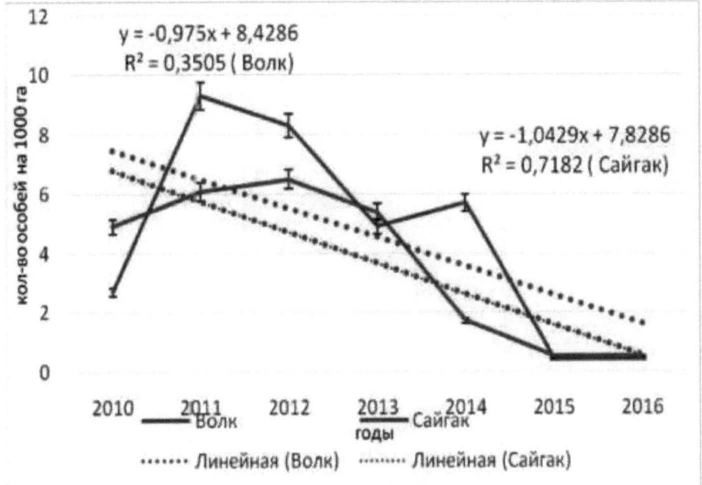

Fig.27. Population dynamics of the wolf population depending on the prey item (saiga antelope) for 2011-2016 (R=0.71).

The dynamics of animal numbers is also characterised by changes in sex composition and age structure in response to periodically changing habitat conditions. Establishing the number of animals and its temporal dynamics over large areas is difficult, and it is not by chance that we do not have any reliable and objective data on the population density of many species, in particular the wolf.

The imperfection of applied methods and organisational difficulties in conducting quantitative surveys over the vast area of the Southern Priaralie have made it difficult to obtain sufficiently accurate estimates of wolf numbers. In such cases, the data are of expert or, as Koli G. (1979) aptly called them, approximate nature. In some cases, a more or less precise idea of the number is necessary, in other cases, a point estimate is sufficient, and in third cases, one can be satisfied only with the knowledge of the general trend of population size: to have some relative indices (indices) objectively reflecting the process of population fluctuation.

However, it is essential that the selected indicators correlate with changes in population size. Such indicators of abundance include traces of animal activity, the number of individuals encountered (or preyed upon) per unit time, etc.

For practical purposes, it is important to have an idea of wolf abundance and its dynamics in certain parts of the range.

For this purpose, it is more convenient to use such an indicator as skins harvested, i.e. to make the information obtained comparable.

The same data can serve as indices of abundance. Comparison of the latter reflects the uneven distribution of this predator in the Southern Priaralie. The obtained data on the harvesting of wolf skins can fairly objectively reflect the state of population size and can even be used to establish certain regularities, as was the case with the lynx and the American squirrel [127, p.161].

In the course of the scientific research we have accumulated extensive materials on the harvesting of wolf skins from 1950 to 2000. However, this material was heterogeneous and unequal; if before the 1990s, the harvest data more or less objectively reflected the population trend, later, due to the disruption of the harvesting mechanism and closure of harvesting offices, these data could no longer serve as reliable indicators. Because of the significant damage caused by the wolf to domestic and wild animals until 1991, the state paid significant bonuses to hunters in addition to the procurement cost of wolf skins to stimulate the extermination of this predator.

According to Reimov R. (2000, 2003) wolf skins were harvested from 1955 to 1970 in the amount of 1,293 skins, averaging 86 skins per year, a harvest rate of 0.57%, and from the 1980s to 1995 averaging 10-15 skins per year (Table 2).

Table 2

Wolf pelt harvest in the Southern Priaralie for 1955-2000
(by Reimov R., 2000,2003)

Name of species	1955-1970.		1980 -2000.	
	Total (pcs.)	Average per year (pcs.)	Total (pcs.)	Average per year (pcs.)
Wolf	1293	86	61	3,1

The graph on the dynamics of harvesting wolf skins shows that in the 80s 41 skins were harvested per year, in the 90s - 17 pieces, by 2000 almost no skins were harvested (Fig.28). The maximum production of these predators occurred in 1955-1970, when the procurement points received 1293 skins, the minimum - in 2000.

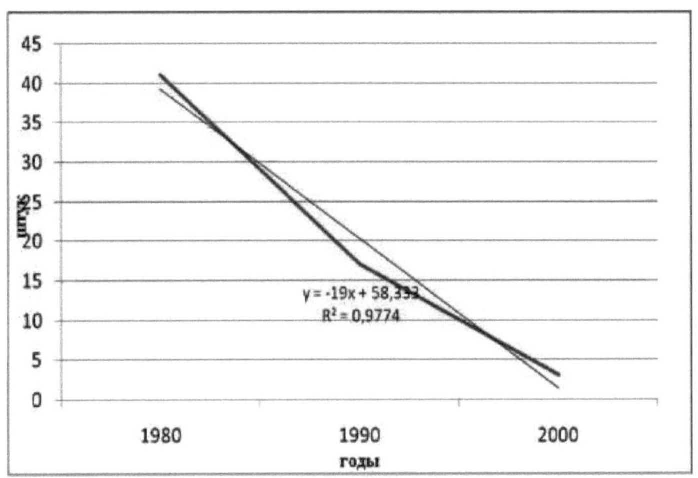

Fig.28. Dynamics of harvesting wolf skins from 1980s to 2000 (according to R. Reimov, 2000,2003).

The results of wolf counting according to the Karakalpak Republican Society of Hunters and Fishermen (2000-2016) are shown in Figure 29. The counts were conducted in the territories of hunting grounds "Dautukul", "Domalak", "Yuzhnoye", Free hunting grounds "Kok tas", "Kok darya", "Lake Baimurat", Hunting farm "Uzyn kair" and hunting grounds "Akpetkei".

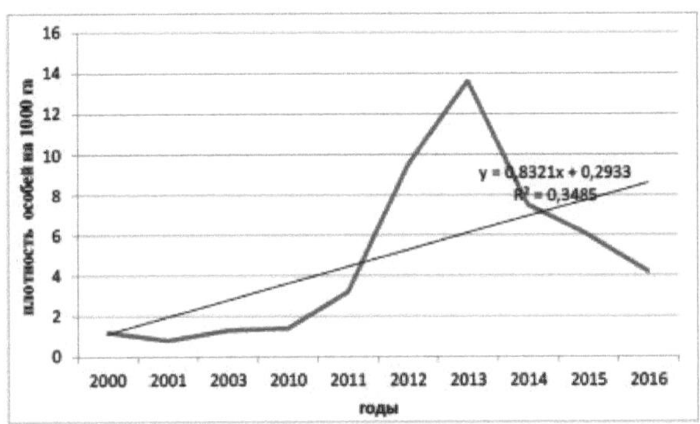

Figure 29. Dynamics of wolf population size according to the Karakalpak branch of the Hunters and Fishermen Society of Uzbekistan (density of individuals per 1000 ha)

Considering the dynamics of the wolf population from 2000 to 2016, it can be noted that the wolf population from 2000 to 2010 tends to decrease, apparently due to the influence of various environmental factors: drying up of lakes, depletion of the forage base, illegal shooting, etc. Since 2010, the wolf population has been gradually increasing, the reason for which is the improvement of protective and forage habitat conditions.

Thus, the results of our studies have shown that under the weak influence of anthropogenic factors, the wolf population size in the Aral Sea region is regulated by variation of animal fecundity and survival rate of young animals, sex composition and age structure of the population, habitat change and mortality from natural causes.

This is consistent with the literature data on the manifestation of these processes, which function as a population mechanism of changes in abundance [127, p.161]. The dynamics of animal numbers is regulated by changes in the sexual composition of the population.

In years of high numbers of animals, males predominate in the populations, and in years of low numbers, females predominate. Changes in the number of predators depend on food availability.

In poorly fed years, most broods are in unfavourable conditions, pups lack food and the weakest die. Analysing the data obtained, we came to the conclusion that predators come into close contact with a variety of animals, especially rodents - carriers of infectious and invasive natural focal diseases.

The regularity of wolf population density distribution, deduced from blanks of wolf skins, can be used in the regulation of wolf numbers.

CHAPTER IV

ECOLOGICAL SIGNIFICANCE AND MANAGEMENT OF
WOLF POPULATIONS
IN THE SOUTHERN ARAL SEA REGION

4.1 Ecological and economic significance

One of the fundamental problems of population ecology, which is of great theoretical and economic importance, is predator-prey relationships. There are no absolutely useful or absolutely harmful predatory animals in nature. Therefore, ecologists, zoologists and hunters of many countries of the world, including the Republic of Uzbekistan, are actively developing methods of identification and quantitative assessment of patterns of relationships in the system "predator - prey", which is of paramount ecological and economic importance. One of the representatives of predators, the wolf, is known to have a certain ecological importance.

Wolves are a major cause of individual mortality of herbivores and play a major role in maintaining their ecological and physiological well-being. One of the aspects of natural selection is realised through wolf predation. In case of complete extermination of the wolf, the growth of ungulate numbers sharply progresses, leading to deterioration of their population and habitat quality. Decreases in weight, linear size and optimal ungulate population structure were particularly common. The spread of various diseases, various natural injuries, physical disfigurement, increased infestation with ecto and endoparasites, etc. have also been observed among ungulates. [127, c.161]. In the subsequent period of the wolf population outbreak, defective ungulate individuals are first of all exterminated. It is at such moments that the "sanitary" role of the wolf is especially manifested. According to the literature data of some authors it is indicated that wolves kill quite healthy, in the prime of life animals. But conducted American scientists' studies show that wolves eat only old and sick victims [147, p.834].

In the course of the study, we took into account two opposing views of experts. Each point of view has its own convincing arguments. The opposing views of experts on the ecological role of wolf predation do not exclude either of them. Most likely, the studies were conducted in different conditions, at different times and reflect the different state of predator and prey populations that do not adequately respond to the wolf.

When assessing the selective role of the wolf in ungulate populations, it is necessary to use both points of view, which we have done. At sharply reduced wolf abundance and disruption of hierarchical structuring in its packs the selectivity of prey increases. Weakened and traumatised victims were observed among their prey. In contrast, at sufficiently high wolf numbers, prey selectivity decreases and elimination becomes more generalised. The above was confirmed on the model wolf population in the Aspantai-Shakaman area of Chimbay district.

There is evidence that if the ungulate population is in poor condition, wolves prey in large numbers on mostly defective individuals. If after some time the condition of the ungulate population improves, quite healthy, strong individuals may be killed by wolves. Especially the sanitary role of wolves is observed in protected areas (reserves, reserves, national parks), because in these optimal ecological conditions with high density of ungulates there is overpopulation, high food competition, depletion of forage resources.

In any ecological conditions, wolves primarily prey on the most vulnerable victims within a population group. Young yearlings are particularly affected, as they are susceptible to foraging stress and any deterioration in habitat conditions makes this age group very accessible to wolves. According to A.A. Sludsky (1981), wolves in Kazakhstan primarily kill young boar and gazelle lambs. Wolves are selective with respect to the age and sex of the prey population and to local environmental conditions. This selectivity depends on temporal and spatial variability.

Considering the predatory role of the wolf from the biocenotic point of view, it should be noted that it is not individual species of prey populations that interact with each other, but their groups, which form relatively homogeneous ecological links. Within each such link there are strongly developed compensatory reactions maintaining functional stability of the system "predators - herbivores - vegetation". Such an ecosystem functions as a unified whole.

When studying the ecological significance of the wolf, it is impossible not to consider its relationship with other predators in the region. There are no large predators competing with wolves in the study area of the Southern Priaralie, but there are some small predators that share a common food base in terms of hares and rodents. These include jackal, fox, corsak.

Systematically, the wolf is most closely related to the jackal, but they have different ecological niches. The jackal mainly lives in dense reed thickets and feeds on carrion left by wolves. There is no competition for shelter. The habitats of jackals and wolves may overlap partially or completely, but they plot their hunting routes differently.

The distribution of jackals depends on the use of the territory by lone or pack wolves. Where non-territorial wolves hunt, the number of jackals is higher than where wolf packs are permanently present. There is evidence that when wolf numbers decrease, jackal numbers decrease. The reason for this is less availability of wolf prey remains for jackals. Based on the above, it can be noted that the biocenotic relationship between wolf and jackal is of the nature of a commodissalism.

Foxes and corsaks are also out of direct competition with wolves due to their food specialisation. Like the jackal, these species can switch to preferential feeding on carrion, in particular the remains of wolf prey (Table 3). According to scientists, when examining the stomachs of foxes, the remains of saigas left behind by wolves were found. A number of kittiwakes were also observed near the remains of wolf prey. During saiga migration, corsaks have been observed chasing herds. However, during migration, most korsaks are killed by wolves [110, p.8-57]. During expeditionary research in 2014, we recorded cases where two adult wolves mauled a jackal on the territory of the tributary farm of the Society of Hunters and Fishermen of the Republic of Karakalpakstan. Bibikov D.I. et al. (1985) described another important ecological benefit of wolves: "It is well known that wolf prey provide for the existence of many "spongers" - small and feathered predators.

Table 3

Biocoenotic relationships of wolves with other predators

Relationship	Types		
	Jackal	Fox	Korsak
Eating the remains of the wolf's prey	+	+	+
The use of dens by wolves		+	
Direct aggression	+	+	+
Presence of prey species and plant food common to wolves in their diet	+	+	+

The economic importance of the wolf lies in how and in what quantity it removes ungulates in forest hunting. Under such conditions, the assessment of its hunting activity becomes negative. The main negative effect of the wolf is as follows: firstly, by increasing in their numbers wolves sharply reduce the number of useful animals, secondly, wolves are carriers of infectious diseases such as scabies and rabies, thirdly, they cause great damage to livestock and local population. Thus, according to our observations, in 2005 in Uchsai-Muynak village of the district there were cases of rabid wolves attacking two residents of the village, after which both died a month later from rabies.

Studies previously conducted in the Southern Priaralie region have established that the assessment of the "harmfulness" or "usefulness" of the wolf was made on single cases, without the necessary study of the whole complex approach to this problem. Harmfulness of the wolf is determined by the fact that it massively destroys domestic animals, which form a significant part of its diet. This is clearly expressed in areas where livestock breeding is well developed (Chimbay, Kegeyli, Takhtakupyr, Karauziak, Muynak, Kungrad districts).

Thus, we found that at sharp increase in the number of wolf population there are concentrations of their packs in densely populated areas, and vice versa - at sharp decrease in the number of wolves there is a selection of individuals specialising on domestic animals.

One of the ecological features of the wolf in the Southern Priaralie is the specialisation of ecological niche use. This means that both at high and low wolf numbers, domestic animals serve as a priority object of their feeding. As a result of the analysis of questionnaire data, the approximate amount of damage caused by wolves to livestock in the territory of the Republic of Karakalpakstan was established (Fig.30).

The analysis showed that the main victims of domestic animals are small and cattle.

67

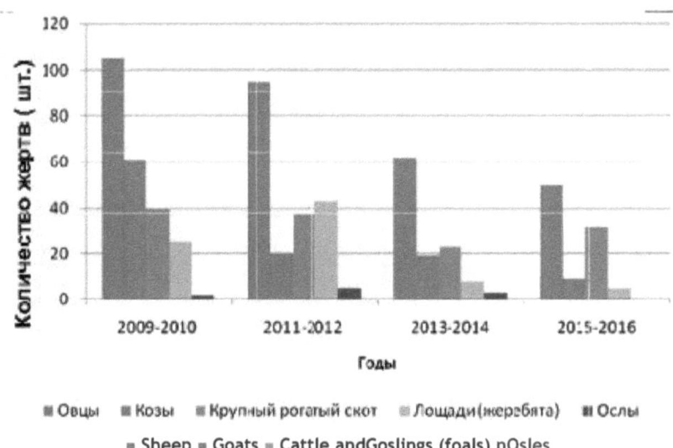

**Fig.30. Dynamics of damage caused by wolves to livestock
in the territory of the Republic of Karakalpakstan in 2009-2016.**

Thus, based on the above, it is clear that the ecological and economic importance of the wolf is that it maintains the number of prey in an optimal state by its predation, significantly affects the ungulate population and can serve as an important regulating ecological mechanism. At the same time, the wolf is a factor regulating interspecific relations among ungulates, and its activity should be evaluated not only from the point of view of its influence on the number of one or another prey species, but also in terms of how reliably and sustainably it protects vegetation from over-utilisation by ungulates. Another equally important significance of the wolf is that its impact transforms the composition of its prey populations and changes the way they use the territory. In this case, it acquires an evolutionary aspect. In any case, the ecological assessment of natural phenomena should not be replaced by the economic one, and these assessments will never coincide [8, p.620]. The wolf is an excellent "breeder", a key species of desert ecosystems. It regulates the number of wild ungulates, preventing excessive population growth and the development of epizootics.

4.2 Management of wolf populations

Until recently, the issue of wolf population management was not as acute as it is today. In the recent past, there was a well-established view on the harmfulness of the wolf and measures to exterminate the predator. On the territory of the CIS countries there was a wolf population eradication struggle. As a result, the wolf was destroyed in most of its range, as well as in the territories of the USA, Mexico and Western Europe. Only since the 30s of the last century scientists began to express their opinion about revision of the entrenched views of the wolf's harmfulness. In the second half of the 20th century, human attitudes towards nature and biodiversity conservation changed radically. Numerous studies of wolf ecology and its relationship with wild ungulates have confirmed the incorrectness of the scientific view of the programme of complete extermination of the predator. The American Society of Zoologists held the First Wolf Symposium in 1967. During this period, various organisations for wolf protection appeared in many countries of the world. At the initiative of WWF (World Wildlife Fund) and IUCN (International Union for Conservation of Nature), great interest was shown in the wolf, and this predator was included in the list of rare and endangered species of the IUCN Red Book, not including Russian subspecies. The IUCN Wolf Specialist Group was organised in 1973. In 1973, the IUCN conference in Stockholm, Sweden, adopted the International Manifesto for Wolf Conservation. This Manifesto includes a Declaration of Principles for Wolf Conservation and recommended Guidelines for wolf conservation. The Manifesto was later revised by the IUCN Wolf Conservation Specialist Group on 31 January 1983, 20 November 1996 and 23 February 2000.

In 1973, the Convention on International Trade in Endangered Species included the wolf in Appendix II (potentially threatened species), except in Bhutan, Pakistan, India and Nepal, where it is listed in Appendix I (species threatened with total extinction). The wolf is also listed in Appendix I (strictly protected species) of the Bern Convention (Convention for the Conservation of European Wild Animals and their Natural Habitats, 19.09.1979). On the basis of this convention, the wolf and its habitat receive full protection, although compliance with this provision is the responsibility of all contracting parties.

It is well known that the main objective of the Bern Convention is to maintain and restore viable wolf populations in coexistence with humans as an integral part of ecosystems and landscapes throughout Europe. As noted in this Convention, the recovery and protection of wolves represent an essential part of the efforts to protect biodiversity in Europe and to ensure the functionality of its ecosystems.

By considering the Berne Convention Plan in relation to the European part, we assume that extending the boundaries to Central Asia will enable a sensible strategy in the management of wolf populations.

Wolves are a rather difficult subject to study for traditional methods, their distribution was limited to remote areas, they were very mobile and their population density was very low. During this time, various methods have been developed to study wolf ecology. In the early 1960s, one method of record keeping was developed - radio tracking. This technique is considered particularly valuable for wolf research. American scientist Kolenoskj (Kolenoskj, 1967) was the first to conduct radio-tracking of wolves in Ontario, a province of Canada. Mech and Frenzel (1971) combined this technology with aerial tracking and observation. After that this method became widely used in many countries of the world. This method is currently being used by Kazakhstan scientists as part of the Altyn Dala project (a large-scale partnership programme of national and international conservation organisations), which has allowed them to determine wolf numbers, their daily and seasonal movements, and their victims.

Wolf management remains an extremely controversial issue due to the complexity of the predator's role in the national economy and wildlife. At present, the Republic of Uzbekistan is one of the countries where the reduction and management of the wolf population remains one of the urgent tasks in the conservation of biodiversity.

The management of wolf populations has its own peculiarities. It is very important to know the spatial territorial structure of family and pack plots. In addition, it is important to have information about the number of indigenous areas occupied by wolves, average fecundity, mortality and age structure of their populations. Knowing the boundaries of indigenous areas of family flocks makes it possible to regulate their numbers more efficiently and competently.

Thus, in order to control the state of wolf resources in the Southern Priaralie region and to coordinate the efforts of inspectors and hunters, it is economically justified to create specialised operational services of wolf management under the controlling bodies, consisting of gamekeepers and gamekeepers. With sparing "regulation" of wolf population and limitation of legal hunting of ungulates it eliminates inter-pack competition of wolf and creates the most favourable conditions for its reproduction.

When solving the problems of wolf population management, it is necessary to implement an integrated approach taking into account ecological-geographical, socio-economic, moral and organisational aspects. Taking into account ecological aspects, wolf population management should take into account the regulation of wolf numbers to an ecologically justified minimum limit, at which viable populations will be preserved and the assessment of damage to the economy will be minimal.

The only alternative approach to wolf management is a differentiated approach. It includes regional and geographical identification of areas where the wolf range should be reduced due to obvious damage to livestock and forestry. Such an approach will allow to maintain the numbers and structure. The wolf population at a certain level for a particular territory. According to D.I. Bibikov (1985) there are 4 types of wolf population control regimes:

1. Strict regulation includes up to habitat reduction in regions with high population density and intensive nature use;

2. Moderate regulation and maintaining the average density at no more than two animals per 1,000 km^2 in regions with low population density and extensive nature management;

3. The status of game animal is in the area of distribution of wild ungulates;

4. Protection - reserves and other protected areas.

The results of the conducted research show that in the conditions of the Southern Priaralie, taking into account the ecological and economic importance of the wolf, it is recommended to apply the category of "moderate regulation" in some places and the category of "protection" in most cases. In these cases, a differentiated approach to the regulation of wolf population size is necessary from the ecological point of view. In our opinion, the presence of wolves may be incompatible with intensive human economic activities, so the best possible integration of human activities with reasonable protection of biodiversity should be planned. Since the regulation of their presence on the territory of a State cannot be entrusted to the individual reaction of herders who suffer great damage or, even more so, of poachers, the Management Plan should establish a series of goals, objectives, criteria and methods on the basis of which the presence of this species will be adjusted. Without limiting the objectives to the protection of viable wolf populations that are not threatened by the most likely risk factors, an initial reflection on the management of wolf populations by zone is needed. This would lead to full protection of wolves only in a certain part of the territory, and thus may provoke conflicts with farmers and shepherds in areas more prone to attacks. This in turn would require a range of measures, both preventive and remedial, including the removal of a few individuals on a local scale.

The feasibility of such an approach should be assessed at ecological, social, administrative, environmental and ethical levels. From an ecological point of view, it seems possible and appropriate: high annual wolf mortality due to illegal shooting occurs mainly in areas of the most highly developed livestock production. Thus, wolf populations can tolerate such sacrifices from poaching if managed properly, paying attention to timing and other criteria.

CONCLUSION

Biological diversity is a great asset for present and future generations. Each species of living things represents a unique result of evolution, which makes the loss of genotypes irreplaceable. Irreversible reduction of biological diversity (at the level of species and ecosystems) can lead to irreversible disturbances in the stability of the entire biosphere.

So far, the regulation of wolf population in the Aral Sea region has been carried out spontaneously, haphazardly, without special consideration of indigenous areas of breeding maternal pairs, without organisation and coordination of efforts of ordinary wolf hunters. The analysis of ecological and ethological peculiarities of territorial groupings of wolf in modern conditions of the Southern Priaralie is the most relevant and timely. Diversity of landscape and climatic conditions, variety of biotopes, different anthropogenic load on wolf population and their habitat determine the formation of those or other features of territorial distribution, movements, daily and seasonal activity, feeding patterns and some other features of wolf ecology in the Southern Priaralie. Quantitative analysis of the diet of the wolf in the Central part of the Southern Priaralie region has shown that the main share belongs to forage resources of anthropogenic origin (about 56%), and the remaining 44% - to the forage of natural origin.

The ratio of different sex and age groups in populations determines the ability to reproduce at a given moment and reveals the trend of population changes. An increase in the proportion of young in populations is the result of favourable breeding conditions, an indicator of population recovery from depression and growth of animal numbers. In years of low numbers of animals, spatial changes in the predator population are observed, manifested in significant separation of individual groups - wolf packs from each other and their confinement to settlements, especially around small lakes. Thus, observation of the sex and age structure of the wolf population should become an important element of prediction of population changes necessary for the rational use of predator resources.

The wolf is characterised by its great ecological plasticity and with a highly developed psyche. This allows it to successfully resist all the variety of ways of fighting it. For a long time, the behaviour of animals, in particular wolves, was defined in terms of their instincts and conditioned reflexes. The results obtained in the course of the study show that the wolf population under study consists of pack-united individuals using certain territories of indigenous sites and solitary animals that are not part of a pack and are characterised by high mobility. The size of the habitat area of the wolf population in the Aral Sea region is determined by a certain landscape and varies in different areas. The size of the territory and the density of one pack depends on the food supply, the availability of shelters and the location of water sources.

Considering the predatory role of the wolf from the biocenotic point of view, it should be noted that it is not individual species of the prey population that interact with each other, but their groups, which form relatively homogeneous ecological links. Within each such link there are strongly developed compensatory reactions that maintain functional stability of the system "predators - herbivores - vegetation". Such an ecosystem functions as a unified whole.

The dynamics of animal numbers is regulated by changes in the sex composition of the population. In years of high numbers of animals, males predominate in the population, and in years of low numbers, females predominate. Changes in the number of predators depend on food availability. In low-feeding years, most broods are in unfavourable conditions, pups lack food and the weakest die. The regularity of wolf population density distribution, derived from the harvesting of wolf skins, can be used to regulate wolf numbers.

When solving the problem of wolf population management, it is necessary to implement an integrated approach taking into account ecological-geographical, socio-economic, moral and organisational aspects. Ecological aspects should be taken into account when managing wolf populations by regulating wolf numbers to an ecologically justified minimum limit, at which viable populations will be preserved and the assessment of damage to the economy will be minimal. The only alternative approach to wolf management is a differentiated approach. It includes regional and geographical identification of areas where the wolf range should be reduced due to obvious damage to livestock and forestry. Such an approach will allow to maintain the wolf population size and structure at a certain level for a particular territory.

As a result of the thesis work on "Ecology of the wolf in the conditions (*Canis lupus* Linnaeus) of the Southern Priaralie" the following conclusions were obtained:

CONCLUSIONS

As a result of research of the scientific work on the topic "Ecology of wolves in the conditions (*Canis lupus* Linnaeus) of the Southern Priaralie" the following conclusions were obtained:

1. Spatial-territorial locations of wolf packs differ in size, the largest territorial areas of packs are typical for the Ustyurt Plateau ($800 - 1200$ km^2) and the smallest - in the central zone of Priaralie ($150 - 220$ km).^{2}

2. For the first time it was revealed that wolves actively use resources of anthropogenic origin: on the Ustyurt Plateau 20%, in the central zone 56%, and feeds of natural origin on the Ustyurt Plateau - up to 80%, in the central zone - up to 44%.

3. In the conditions of the Southern Priaralie, 2 main types of lairs have been identified: formed and weakly formed dens in protected places, which have transitional forms distributed across different landscapes. Wolves make lairs in formed dens on the Ustyurt Plateau, riparian and reed thickets, Kyzylkum desert zone, and weakly formed dens are inherent to wolves in the Lower Amudarya River.

4. Gonation in wolves in the Southern Priaralie occurs later than in other areas of its distribution, starting in early February and ending in the second decade of February. The number of pups in a brood varies from 2 to 10, with an average brood size of 5.9 wolves. The fecundity of females depends primarily on the availability of food during the breeding season, as well as the age of the she-wolf, the density of its own population and the intensity of human extermination.

5. On the basis of the obtained data, the sex ratio ($:$) in the model population of wolves in the conditions of the Southern Priaralie was determined, where a slight predominance of males (5:3) was observed. Due to intensive poaching by shepherds and local population, there was a sharp increase in the proportion of females in the offspring (1:2.5).

6. Saiga ($R=0.71$) and gazelle ($R=0.39$) are the main prey items of the wolf, which depend linearly on predator density. This shows that the wolf's relationship with these prey items has a positive trend.

7. The ecological and economic importance of wolves lies in the fact that they, through their predation, maintain the number of prey in an optimal state, significantly influence the ungulate population and can serve as an important regulating ecological mechanism. The relationship of the wolf with other predatory mammals of the biocenosis will lead to their partial overlap of spatial and foraging eco-niche.

8. The need for a differentiated approach to the management of wolf populations in natural zones of the Southern Priaralie depends on the ecological features of territorial groupings of wolves, as well as on the anthropogenic transformation of landscapes and on the regime of nature management.

9. Taking into account the ecological and economic importance of wolves in the conditions of the Southern Priaralie, it is recommended to use the category of "moderate regulation" and in most cases the category of "protection" to regulate the number of wolf population.

10. The developed GIS ecological maps of permanent wolf family plots are recommended for monitoring the status of wolf fecundity, mortality, population structure, and forage resources.

PRACTICAL RECOMMENDATIONS

1. It is necessary to monitor the spatial distribution of the wolf population in the Southern Priaralie in order to collect information on the number of indigenous areas occupied by it, average fecundity, mortality and age structure of its populations. Knowledge of the boundaries of indigenous areas of family packs makes it possible to regulate their numbers more efficiently and competently.

2. To control the state of wolf resources in the Southern Priaralie region and to coordinate the efforts of inspectors and hunters, it is economically justified to create specialised operational services of wolf management under the controlling bodies, consisting of gamekeepers and gamekeepers.

3. The control of wolf numbers and the reliability of counting, spatial distribution of the wolf is possible only with ecological mapping of permanent family packs of the wolf population, with annual monitoring of the state of fecundity, mortality, population structure and the state of forage resources.

4. Taking into account ecological aspects, the management of wolf populations should take into account the regulation of wolf numbers to an ecologically justified minimum limit at which viable populations will be preserved and damage to the economy will be minimised .

LIST OF REFERENCES

1. Aimuratov R.P., Pirzhanova R.K. Dynamics of the flora of Karakalpakstan // "Dynamics and potential of the natural environment of Karakalpakstan. - Nukus. - Karakalpakstan. - 2017. - C.78 - 90.

2. Badridze Ya. Wolf Issues of ontogenesis of behaviour, problems and method of introduction. - M: Izd - voe Geos. - 2004. -34 c.

3. Barabash - Nikiforov, Formozov A.N.-Teriology. - M: Gosizdat. - 1963. - 396 c.

4. Batyrov B.H. To the history of the theriofauna of the Zerafshan valley in the Anthropogeny. // Materials of the Republican scientific and technical conference of young scientists and graduate students. - Samarkand: 1968. - C.60 - 63.

5. Bakhiev A., Butov K.N., Tadjitdinov M.T. Dynamics of plant communities of the southern Aral Sea region in connection with changes in the Aral Sea basin hydroregime. - Tashkent. - Fan. - 1977. - 80 c.

6. Bakhiev A.B., Treshkin S.E., Kuzmina J.V. Analysis and assessment of the impact of anthropogenic factors on the riparian vegetation of the Amu Darya delta // Bulletin of KKO AS of the Republic of Uzbekistan. - № 2. -Nukus. - 1996. - C. 3 - 8.

7. Bykova E.A., Esipov A.V. Saiga in Uzbekistan: trends in status change, ecology and protection measures // Proceedings of the XXIX International Congress of Biologists and Hunting Scientists. - M. - P.8 - 9.

8. Bibikov D.I. et al. Wolf. - M.- Nauka. -1985. - 605 c.

9. Bologov V. Control of wolf numbers // Hunting and hunting economy. - 1984. - № 2. - C.4 - 5.

10. Bondarev A.Ya. Sex ratio in wolves is an indicator of population fluctuations // Ecological bases of protection and rational use of predatory mammals: Proceedings of the All-Union Council. - Moscow: Nauka. - 1979. - 88 c.

11. Bondarev A.Ya. To characterise the reproduction of wolves in Altai and southern Western Siberia // Quantitative methods in vertebrate ecology. - Sverdlovsk. - 1983. - C.78 - 91.

12. Bondarev A.Ya. Wolf of the south of Western Siberia and Altai. -Barnaul. - Izd-vo Barnaulsk. gos. ped. un-ta. - 2002. - 176 c.

13. Bondarev A.Ya. On the principles of regulation of wolf numbers // Bulletin of Altai State Agrarian University. - №9 (95). - 2012. - C. 70 -71.

14. Boreyko V.E. In Defence of Wolves. - Protection of wildlife. - Vyp. 68- Kiev. - 2011. - 156 c.

15. Becker L., Nasiadka P., Korablev P.N., Bolotov V.V.. Impact of wolf on human activity and limits of human regulation of wolf numbers in the Tver region (Russian Federation) // XXIX International Congress of Biologists - Hunting Scientists. - M.- 2009. -C.5 - 6

16. Vanisova E.A. Attractors in the biological signalling field of some mammalian species. -Autoref. diss.... Candidate of Biological Sciences - Moscow - 2013. - 25 c.

17. Viktorov S.V. Ustyurt desert and issues of its development. - Moscow: Nauka. - 1971. -177c.

18. Geptner V.G. Mammals of the Soviet Union. - T.2. 4.1. - Sea cows and carnivores. - Moscow: Vysh. shk. - 1967. - C.123 - 193.

19. Gubar Yu.P. Methodical recommendations on wolf counting by the method of habitat mapping / Central Scientific Research Institute of Glavohoty RSFSR. - M.- 1987. - 29 c.

20. Gursky I.G. Wolf in the North-Western Black Sea coast (habitat, population structure, reproduction) // Bul. of MOIP, Biol. MOIP, Department of Biol. - 1978. - T.83. - Vol. 83. 3. - C. 29-38.

21. Dewsbury D. Animal behaviour: Comparative aspects. - ML: World. -1981. - 40 c.

22. Zavatsky B.P. To the ecology of the wolf (Canis lupus L.) of the Western Sayan // Hunting - commercial resources of Siberia. Novosibirsk: Nauka. - 1986. - C. 118 - 125.

23. Zavatsky B.P. Methodical guidelines for wolf counting by the method of habitat mapping. - M. - 1987. - 29 c.

24. Zvorykin N.A. Wolf. - M.-L.; KOIZ. - 1937. - 119 c.

25. Zorina Z.A., Poletaeva I.I. Elementary thinking of animals. / Study guide. - Moscow: Aspect Press. - 2002. - 320 c.

26. Zyryanov A. N. Wolf in the Stolby Nature Reserve // Bulletin. MOIP. Biol. 1995. T. 100. - Vol. 1. - C. 20-33.

27. Karpenko N. T. Ecology of the wolf //Astrakhansky Vestnik of Environmental Education. -№ 2 (18). -2011. - C. 165 - 167.

28. Klevezel G.A. Principles and determinations of the age of mammals M: T - wo scientific editions of KMK. - 2007. - 283 c.

29. Kleimenova I . E. Ecological-geographical zoning Karakalpak Ustyurt // OGU Bulletin No. 10 (116). - 2010. - C.106 -111.

30. Kozlov V. V. Wolf and ways of its extermination. - M. Selkhozgiz, - 1955. -85 c.

31. Kozlov V.V. Wolves of the forest-steppes of Siberia and their extermination. - Krasnoyarsk: Kn. izd - vo. 1966. - 129 c.

32. Kozlovsky I. Regulation of wolf numbers - the responsibility of the state// Hunting and hunting economy. - 2015. - № 5. - C. 1 -5.

33. Kohli G. Analyses of vertebrate populations. - M. Mir. - 1979. - 362 c.

34. Korytin S.A. Trudy VNIIOZ a.- Kirov. - 1976. - 500 c.

35. Korytin S.A. Smells in the life of animals. - MOSCOW: URSS. - 2010. -127 c.

36. Korytin S.A. Comparative Behaviour of Dogs (*Canidae*). - Kirov: VNIIOZ. -2011. - 64 c.

37. Kochetkov V.V. Biology of the wolf in the Upper Volga region (on the example of the area of the Central Forest State Reserve). - autoref. diss. Candidate of Biological Sciences. - M., 1988. - 20 c.

38. Kochetkov V. Management of wolf populations: desired or actual// Hunting and hunting economy. - 2013. - № 2. - C. 10 -15

39. Krushinsky L. V. Biological bases of reasoning activity. Moscow: Izd-vo MSU. - 1986. - 105 c.

40. Kudaktin A.N. Wolf behaviour in a protected ecosystem // Wolf behaviour / IEMEJ of the USSR Academy of Sciences. - 1980. - C.90 - 102.

41. Kudaktin A.N. Wolf of the Western Caucasus   (Ecology, behaviour, biocenotic position). - autoref. dis...kand. biol. sciences. - M.- 1982. -22 c.

42. Kudaktin A.N. Wolf in the Russian Caucasus - management strategy // XXIX International Congress of Biologists - Hunting Scientists. - M.- 2009. - C.5.

43. Kudaktin A.N. Modern distribution and abundance of wolf on the territory of the Southern Federal District // Herald of Hunting Science. - 2011. - Vol. 8. - № 1. - C. 26 -34.

44. Kudaktin A.N. Methods of predator counting in the Caucasus Mountains // Vestnik Okhotovedeniya. - Vol. 8. -№1. -2 011. - C.104 - 108.

45. Kidirbaeva A.Y., Mamabetullaeva S.M. Ecology of the wolf (CanisLupusLinnaeus,1758) in modern conditions of the Southern Priaralie // Izvestiya Dagestan State Pedagogical University. - Series Natural and exact sciences. - Dagestan. -2009. - №2 (7). - C.41- 43.

46. Kidirbaeva A.Y., Mambetullaeva S.M. To the question of determining the population size of the wolf in the conditions of the Southern Aral Sea region // Bulletin of the KCO of the Academy of Sciences of the Republic of Uzbekistan. - Nukus. - 2009. -№ 4. - C.44 - 45.

47. Kidirbaeva A.Y., Mambetullaeva S.M. Ecological structure of the wolf population in the conditions of the Southern Priaralie // Uzbek Biological Journal.- 2011.- № 4. - C.38 - 41.

48. Kidirbaeva A.Y., Mambetullaeva S.M. Predatory mammals in the transformation of the natural environment of the Southern Priaralie // Reports of the Academy of Sciences of the Republic of Uzbekistan. -2014. - №4. - C.80 - 81.

49. Kidirbaeva A.Y.Yu.Pack and territoriality of the wolf in the conditions of the Southern Priaralie // Bulletin of the KKO of the Academy of Sciences of the Republic of Uzbekistan. - 2015. - №4. - C.63 - 66.

50. Kidirbaeva A.Y. Analysing the dynamics of wolf population dynamics in the conditions of the Southern Priaralie // UzMU Khabarlari. -2016. - №3/2. - C.52 - 55.

51. Kidirbaeva A.Yu. Burrows and shelter of predatory mammals in the conditions of the Southern Priaralie// Vestnik KKO AS RUz. - 2017. - №1. - C.65 - 68.

52. Kidirbaeva A   .Yu.   Nutrition of wolf in the conditions of Southern Russia Priaralie// Vestnik KKO of the Academy of Sciences of the Republic of Uzbekistan. - 2017. - №2. - C.17 - 19.

53. Kidirbaeva A.Y., Mambetullaeva S.M., Asenov G. To the ecology of the wolf (*Canis lupus Linnaeus,1758*) in the conditions of the Southern Priaralie // Proceedings of the Republican scientific-practical conference "Problems of rational use of natural resources of the Southern Priaralie". - Nukus. - 2008. - C.68 - 69.

54. Kidirbaeva A.Y., Mambetullaeva S.M. Some regional peculiarities of wolf ecology in the Southern Priaralie // Proceedings of the scientific conference "Actual problems of biodiversity conservation". - Tashkent. - 2010. - C.53 - 56.

55. Kidirbaeva A.Y., Mambetullaeva S.M. Some features of the bioecology of the wolf in the Southern Priaralie // Proceedings of the Republican Scientific and Practical Conference "Achievements, prospects of development and problems of natural science". - Nukus. - KSU. - 2011. - C.21.

56. Kidirbaeva S.M., Mambetullaeva S.M. Features of the population structure of the wolf in the conditions of the Southern Aral Sea region // Proceedings of the Republican Scientific and Practical Conference "The role of women scientists in the development of society. - Nukus. -KGU. - C. 164 - 167.

57. Kidirbaeva A.Yu. Problems of wolf population management in the Southern Priaralie// Proceedings of the International Conference "Sustainable Development of the Southern Priaralie". - Nukus. - 2011. - C. 39 - 40.

58. Kidirbaeva A.   Y., Mambetullaeva S.M.,   Matmuratov M.   . Ecological and economic significance of the wolf in the Southern Priaralie // Proceedings of the IV International Scientific and Practical Conference "Problems of rational use and protection of bioresources of the Southern Priaralie, Nukus, Ilim, 2012. - C.75 - 76.

59. Kidirbaeva A.Y., Matsapaeva I., Nysanova S. Biological basis of wolf accounting by mapping method // Proceedings of the Republican Scientific and Practical Conference "Rational use of natural resources of the Southern Priaralie". - Nukus. - 2012 - C.64 - 65.

60. Kidirbaeva A.Yu. To the question of systematics of the wolf of the Southern Priaralie // Materials of the Republican scientific - practical conference "Rational use of natural resources of the Southern Priaralie". - Nukus. -2013. - C.66 - 69.

61. Kidirbaeva A.Y. Predatory mammals of the Southern Priaralie // Proceedings of the Republican scientific-practical conference "Ecology and Physics". - NSPI. - Nukus. - 2013. - C.73.

62. Kidirbaeva A.Yu. Packs and territorial movements of wolves in the conditions of the Southern Priaralie// Proceedings of the Republican Scientific and Practical Conference "XXI esir - intellectual jaslar esiri". - 2015. - C.152 - 155.

63. Kidirbaeva A.Yu.Theory of biological signal field in the study of ethological aspects of predatory mammals of the Aral Sea region // Materials of the Republican scientific-practical conference "IV Rational use of natural resources of the Southern Aral Sea region". - Nukus. - 2015. - C. 143.

64. Kidirbaeva A.Y. Predatory mammals in the transformation of the natural environment of the Southern Priaralie// Proceedings of the XXIV International Scientific and Practical Conference "Science in the Modern World". - (Russia, Taganrog). -2015. - C.6-9.

65. Kidirbaeva A.Yu. To the issue of wolf population management in the Southern Priaralie// Proceedings of XXIV International Scientific and Practical Conference "Science in the Modern World". - (Russia, Taganrog). - 2015. - C.9 - 12.

66. Kidirbaeva A.Yu. Some aspects of wolf ecology in the conditions of the Southern Priaralie// Proceedings of the Republican Scientific and Scientific-Technical

Conference "Uzbekistonning bioecologik muammolari". - Termez. - 2016 - C.19 -20.

67. Kidirbaeva A.Yu. The current state of the wolf population in the South Prearalie // Proceedings of the Republican Scientific and Practical Conference "V Rational use of natural resources in the South Prearalie". - Nukus. - KSU. -2016. - C.51-53.

68. Kidirbaeva A.Yu. Some features of wolf reproduction in the conditions of the Southern Priaralie // Proceedings of the Republican Scientific and Practical Conference "VI Rational use of natural resources in the Southern Priaralie". - Nukus. - KSU. -2017. - C.51 - 54.

69. Kohli G. Analyses of vertebrate populations. - M.: Mir. - 1979. - 363 c.

70. Lavov M.A. Numbers and peculiarities of the way of life by regions. Krasnoyarsk Krai, Irkutsk and Chita Oblasts / Wolf. Origin, systematics, morphology, ecology. -M.: Nauka, 1985. - C. 539 - 543.

71. Lakin, G.F. Biometrics. - Moscow: Vyshaya Shkola, 1980. - 292 c.

72. Lineitsev S. N. Wolves of the Putorana Plateau // Hunting and hunting economy. 1983. - № 3. - C. 7 - 8.

73. Lopatin G.V., Dengina R.S., Egorov V.V. The Amu Darya Delta. - M., L: Izd-constituion of the USSR Academy of Sciences. -1958. - 148c.

74. Manteifel P.A. Life of fur-bearing animals. - M.- 1947. -55c.

75. Makridin V.P. Wolf // Large predators and ungulate animals. - M.: Lesn. pro-mst. -1978. - C. 8 - 50.

76. Mac - farland Animal behaviour: psychobiology, ethology and evolution - M.: Mir, 1988. -520 c.

77. Marakov S.V. In the jungles of Pribalkhashia. - M: Nauka. -1969. - 25c.

78. Mambetullaeva S.M., Analysis of the dynamics of the number of commercial populations of muskrat. //Autoref. dissertation...kand.biol.nauk. - Ekaterinburg. - 1994. - 24c.

79. Mambetullaenva S., Reimov R. Study of the correlation between the influence of abiotic and biotic factors on the population dynamics of the Zakaspian vole // Bulletin of the KKO of the Academy of Sciences of the Republic of Uzbekistan-1997. - № 4. - C.80-83.

80. Mambetullaeva S.M., Kidirbaeva A.Y. Bioecological features of the wolf (CanisLupusLinnaeus, 1758) in the conditions of the Southern Aral Sea region // Scientific Medical Bulletin. -Tambov (Russia). - 2015. - №2(2). - C.76 - 82.

81. Menning O. Animal Behaviour: An Introductory Course. - M.: Mir. -1982. - 100c.

82. Mikhailov, A. Mamcntov A. // The wolf is a national problem // Hunting and hunting economy. - 2004. - N 7. - C. 10 -12.

83. Mozgovoy D.P., Vladimirova E.D. Signal fields and animal behaviour in signal-information environment // Izvestia Samara Scientific Centre of the Russian Academy of Sciences. - т.4. - №2. - 2002. - C.207-215

84. Momotov I.F. Plant complex of Ustyurt. -Tashkent: Izd. of the Academy of Sciences of Uz. SSR. - 1953. - 200 c.

85. Novikov G.D.   Field studies on the ecology of terrestrial animals

Vertebrates - M.: Sov. Nauka. -1953. - C.187 - 263.

86. Novikov G.A. Predatory mammals of the fauna of the USSR Y. - M.; L.: Izd - vo AS USSR. - 1956. - 293 c.

87. Nikolsky A.A., Rozhnov V.V. Biological signalling field of mammals. - M.- Partnership of scientific editions of KMK. - 2013. - 323 c.

88. Nuratdinov T. Ecology of predatory animals of the Amudarya River and their economic importance. // autoref.dis... Candidate of Biological Sciences. - Alma-Ata. - 1969. - 21c.

89. Ovsyanikov N. G. Behavioural mechanisms of intrapopulation processes of predatory mammals of the Arctic // authroref. diss.d.b.n.s. - Moscow. - 2012. - 48 c.

90. Oreshin A.M. Hydrological sketch of the Amu Darya section from Tuyamuyun to the Aral Sea. - In: Materials on productive forces of the UzSSR. -Vyp 10. -Tashkent. Publishing House of the USSR Academy of Sciences. - 1959. - C.35-53.

91. Pavlov M.P. Wolf. - M.- Agropromizdat. - 1990. - 351c.

92. Pavlinov I.Ya. Systematics of modern mammals. -Moscow University. - 2003. -C. 196-197.

93. Palvanazov M. Predatory animals of Central Asian deserts. - Karakalpakstan, 1974. -320 c.

94. Palvaniyazov M. Mammals of the Southern Priaralie under conditions of anthropogenic landscape change - Tashkent. - FAN. -1990. -76c.

95. Plaksa S. A., Yarovenko Y. A., Mirzoev G. A. Z. Spatial distribution and dynamics of wolf numbers in Dagestan // Izvestia Dagestan State Pedagogical University. - Ser.: Natural and exact sciences. - 2011. - № 3. - C. 47-53.

96. Plokhinsky N. A. Biometrics. - Novosibirsk: Izd vo SO AS USSR. - 1961. - 364 c.

97. Plokhinsky, H.A. Dispersion analysis of the power of influence // New in biometrics. - Moscow: MGU ed. -1970. - C. 31- 67.

98. Priklonsky S.G. Extent of daily movement and some issues of ecology and importance of the wolf in winter // Proc. of the Central Research Institute of Glavohoty RSFSR. Central Research Institute of Glavohoty RSFSR. -Winter route registration of hunting animals. - M., 1983. - C.131-158.

99. Reimov R. Mammals of the Southern Priaralie (ecology, protection and utilisation). - Tashkent: Fan. - 1985. - 96 c.

100. Reimov R. On the state and tasks of fauna protection in Karakalpakstan // Bulletin of KKO AS RUz. - 2000 г. №1-2. - C.3-6.

101. Reimov R.R., Reimov A.R. Ecological aspects of protection and utilisation of the population of terrestrial vertebrates of the Aral Sea region under the conditions of the Aral crisis // Bulletin of the KKO of the Academy of Sciences of the Republic of Uzbekistan. - 2000 -№1-2. -C.9 - 11.

102. Reimov R. Condition of mammals of the Khorezm oasis under conditions of anthropogenic transformation of the landscape // Bulletin of the KKO of the Academy of Sciences of the Republic of Uzbekistan. - 2000 г. №1-2. - C.3 - 6.

103.   Reimov R.R., Reimov A.R. Hunting animals of the Khorezm oasis and adjacent deserts some aspects of their conservation and sustainable use // Bulletin of the KCO of the Academy of Sciences of the Republic of Uzbekistan. -2003 г. №3. - С.30 -33.

104.   Reimov R., Pirzhanova R., Tazhimuratov P. Regularities of formation of faunistic composition of the Southern part of Karakalpakstan // Bulletin of KKO AS RUz. -2005 г. №5. - С.17-20.

105.   Reimov R., Pirzhanova R., Tazhimuratov P. Peculiarities of faunistic composition of the Central part of Karakalpak Ustyurt // Bulletin of KK FAN UzR. - 2006. №1. - С.12-15.

106.   Rogov M.M. Hydrology of the Amu Darya Delta. - L.: Gidrometizdat. - 1957. - С. 10 - 17.

107.   Rukovsky, N.N. Sanctuary of quadrupeds. - M. "Agropromizdat". -1991. - С.84 -109.

108.   Ryabov L. Wolf: origin, systematics, morphology, ecology. // "Hunting and hunting economy. -№ 3. - 1980. - С.12 - 14.

109.   Sludsky A. A. Relationship of predators and prey (on the example of antelopes and other animals and their enemies) // Proc. of the Institute of Zoology. Academy of Sciences of the KazSSR. Alma-Ata, 1962. - T. 17. - С.24 - 143.

110.   Sludsky A.A. Wolf (*Canis lupus.,* 1758) Mammals of Kazakhstan. - Alma-Ata: Nauka. -1981. - T. 3. - Ч. 1. - С.8-57.

111.   Smirnov B.C. Control of wolf population dynamics by age composition of prey animals (methodical recommendations). - Sverdlovsk: UNTs of the Academy of Sciences of the USSR. -1985. -75 с.

112.   Smirnov M.N. Wolf (*Canis lupus* L). -Mammals of the Lake Baikal basin. - Novosibirsk: Nauka. -1984. - С. 44 - 45.

113.   Suvorov A. P. Wolf and ungulates: management facets // Hunting and hunting economy. -2004. - № 3. - С. 1-3.

114.   Suvorov A. P. Wolf in the Yenisei basin (biological aspects of population management) // autoref. diss. Candidate of Biological Sciences - Krasnoyarsk -2004. - 27 с.

115.   Suvorov A.P. Wolf: problems of resource management // National Hunting Journal. -№9. - 2009. - С.18 - 22.

116.   Suvorov, A. Wolf: from extermination to population management // Hunting and hunting economy. - 2011. - № 12. - С. 1-3.

117.   Suvorov A. Wolf counting by occupied family plots // Hunting and hunting economy. - 2014. - № 11. - С. 12-14

118.   Suchkov S. P. Soils of Khorezm region. - In the book of Materials on productive forces of Uzbekistan, vol.10. Natural conditions and resources of the lower reaches of the Amu Darya. -Tashkent. - Iz-voy AN UzSSR. -1959. - С.202 - 209.

119.   Tazhimuratov P.A. Modern floristic composition of the Karakalpak part of Ustyurt. // Bulletin of KKO AS RUz. -2001. - №6. - С. 20 - 21.

120.   Taubaev T. Aquatic vegetation of the Amu Darya lower reaches. Author's thesis

of Candidate of Biological Sciences. - Tashkent. -1954. -17c.

121.   Tirronen K. F. Large predatory mammals of the Karelian-Murmansk region (ecology, management, protection). - 03.00.16 - ecology. - 06.02.03 - animal breeding and hunting science. - autoref. diss. k.b.n.- Balashikha. - 2009. - 12 c.

122.   Tumanov I.L. Biological features of predatory mammals of Russia. -SP. - Nauka. -2003. - C. 52-57.

123.   Tupikova N.V., Komarova L.V. Principles and methods of zoological mapping. - M: Iz-vo Moscow University. - 1979. - 192c.

124.   Fedosenko A.K. Wolves. - Alma-Ata. -Kainar. -1986. - 95 c.

125.   Filimonov A.N. Biology of the wolf of the semi-deserts of Kazakhstan: Cand. diss. - M. -1982. -27c.

126.   Filimonov A.N. Observations on wolf and saiga in the south of Aktobe region / Ungulate faunas of the USSR. - M.: Nauka. -1975. - C.207-208.

127.   Filonov K.P. Predatory mammals. - M.-1981. - 161c.

128.   Tsyndyzhapova S.D. State of the wolf population and its control in the Irkutsk region // Materials of the conference dedicated to the 50th anniversary of the Faculty of Hunting Science. Ч. 1. - Irkutsk: Irkutsk State Hunting Academy -2000. - C.209 - 215.

129.   Tsyndyzhapova S.D. Specialisation of wolf feeding in Pribaikalye // Conference of faculty and postgraduate students 29 February 3 March 2000. - Irkutsk. - 2000. - C.39.

130.   Tsyndyzhapova S.D. Ecology of the wolf (Canis lupus L., 1758) in the conditions of specially protected natural areas (on the example of the Pribaikalsky National Park). - Ulan-Ude, 2003. - 20 c.

131.   Shkvyrya G.M. Distribution, peculiarities of ecology and behaviour of the wolf *(Canis lupus)* on the territory of Ukraine. - Kiev. -2008. -28 c.

132.   Shkvyrya M.G. Human-predator conflict on the territory of Ukraine. - K.: Print Quick. - 2012. -72 c.

133.   Chauvin R. Animal Behaviour. - Moscow: KD "Librocom". - 2009. - 487 c.

134.   Hernández - Blanco José Antonio The spatial and ethological organisation of minimal population groupings of the wolf Canis lupus lupus L., 1758 in a comparative aspect. - 03.00.08 - zoology. - autoref. diss.... Candidate of Biological Sciences - Moscow. -2004. - C.25.

135.   Ehrman D., Parsons P. Genetics of Behaviour and Evolution. - M.: Mir. - 1984. - 568c.

136.   Ozolins J., Zunna A., Pupila A., Ornisats A., Bagrave G. Changes in nutrition, demographic structure and reproduction of wolf and lynx in Latvia associated with the recent implementation of conservation policy // XXIX International Congress of Biologists - Hunting Scientists. - M.- 2009. - C.5.

137.   Kharchenko N.N. Animal burrows, their structure, functions, typology // Forest Bulletin. - №2. - 2005. - C. 72-84

138.   Hynde R. Animal Behaviour. - M.: Mir. -1975. - 856 c.

139. Hedrzak M., Cywicka D., Grzygorzyk O. Analysis of damage caused by wolves (Canislupus) in south-west Poland // XXIX International Congress of Biologists - Hunting Scientists. - M.-2009. - C.5.

140. Cherenkov S.E., Poyarkov A.D. Wolf, jackal. - M. - Astrel Publishing House. - 2003. - C.7- 95.

141. Corbet G. B. The mammals of the palaearctic region. A taxonomic review British Museum (Natural History) and Cornell University Press, London and Ithaca (NY) vii 1978. -314pp.

142. Ellerman, J.R.; Morrisson-Scott,T.CS. Checklist of Palearctic and Indian mammals. Second edition. British Museum (Natural History). - London. -1966. - p. 1756 -1946.

143. Goldman E. The wolves of Norts America: Classification of wolves // In: The Amer. Wildlife Inst Wash. -1944. - Pt.2. - p.387-636.

144. Haber G, 1996. Biological, conservation and ethical implications of exploiting and controlling wolves // Conservation biology. - V. 10, № 4. - p. 1068 - 1081.

145. Kidirbaeva A.Yu., Mambetullaeva S.M. Ethological aspects of the wolf *(Canis lupus linneus, 1758)* in the aral sea region // Austrian Jurnal of Technical and Natural Sciences .- 2016.- № 11.-12.-p.3-5.

146. Kolenoskj G. Wolf predation on wintering deer in east-central Ontario // J. Wildlife Manag. Wildlife Manag. -1972. - Vol. 36. - № 2: - p. 358 - 359.

147. Mech D. The wolf: the ecology and behaviour of an endangered species// The Natur. Hist. Garden City Press. -1970. - 834 p.

148. Mech D.L., Frenzel L. Ecological studies of the timber wolf in Northeastern Minnesota -US Dep. of Agr. Agr. Forest.Serv. Res.Pap.1971 a NO-52. - 62

149. Mech D. Canis lupus: Mammalian species // Amer. Soc.Mammal. -1974. - p 6.

150. Mech D. Wolf pack buffer zones as prey reservoirs // Science. -1977. -Vol. 198. -№ 4314. -p. 320-321.

151. Mech D. L., Boitani Wolves: Behaviour, Ecology, and Conservation University of Chicago Press. -2003. - 472 p.

152. Murray D. L., Waits Received L. P. Taxonomic status and conservation strategy of the endangered red wolf: a response to Kyle et al. (2006). //Conserv Genet: Springer Science+Business Media B.V. 2007. -№8. -483-485p.

153. Peterson R Wolf ecology and prey relationship on Isle Royale // U.S. Nat. Park Serv. Sci. Monogr. Ser. 1977. -№ 71. -p. 210.

154. Pocock R. The races of Canis lupus. - Proc.Zool. Soc.London. - 1935. - p.647-686

155. Rauh R.A. Some aspects of the population ecology of wolves, Alaska /R.A. Rauh // Amer. Zool. 1967. -№ 7. -p. 253-266.

156. Reinhardt Ilka, Kluth Gesa, Nowak Sabina, Myslajek Robert W.Standards for the monitoring of the Central European wolf population in Germany and Poland // BfN -Skripten 398. - 2015. - 44p.

Figure 1. Typical wolf habitats
(Aspantai - Shakman model site)

Figure 2. Measurements of wolf tracks according to the method of Novikov (1953)

Figure 3. Detected wolf burrow in the territory of the Aspantai-Shakaman model site

Figure 4. Wolf excrement found in the gregarious area

Figure 5. Conducting a survey among shepherds

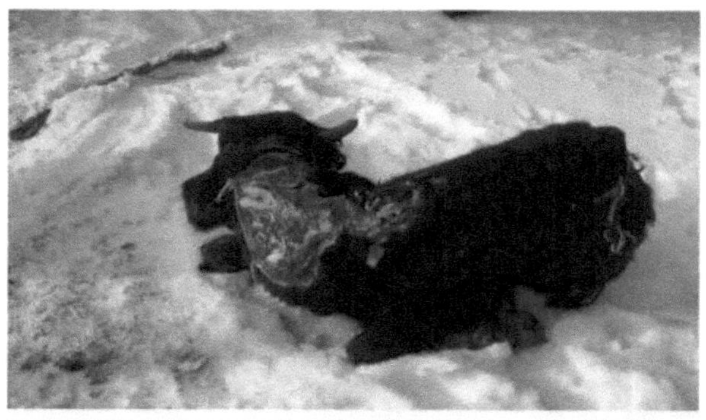

Figure 6. Domestic animals eaten by wolves

ANKETA NO.

1. FULL NAME
2. Location of the questionnaire
3. Place of work
4. Place and time of the wolf encounter (year, date, month)
5. Number of wolves encountered
6. Body size of wolves encountered

Large

Medium

Shallow

7. The location of the lair
8. Number of pups in the den
9. Incidents of livestock attacks (place and time)
10. Incidents of attacks on people (place and time)

(A) A wolf with rabies

11. Consequences of wolf attacks

(A) Remains of cattle carcasses(pcs.)

B) injuries to people

12. Incidents of dog attacks
13. Determining the number of wolves by "howling"
14. Occurrence of wolf tracks and tracks of life activity

A) traces

B) faeces

B) food residues

15. If your observations were during the expedition, please indicate the route (direction), place and time.

Completion Date :

Caption:

WOLF RECORD CARD

Location name

Executors

Date Hours of operation

Weather conditions

Route travelled (km)

Occurrence of wolves (number)

Definitions of wolf by voice

The remains of wolf food eaten

Author information

Arzygul Yuldashevna Kidirbayeva

Biologist - hunter, Doctor of Philosophy in Biological Sciences (PhD), Associate Professor at the Department of Ecology and Soil Science of Berdakh Karakalpak State University. Author of more than 100 scientific works devoted to the ecology of predatory mammals, problems of nature protection and ecosystems of the Southern Priaralie.

Printed by Books on Demand GmbH, Norderstedt / Germany